ENVIRONMENTAL IMPACT STATEMENTS

Second Edition

ENVIRONMENTAL IMPACT STATEMENTS

Second Edition

Jacob I. Bregman

CRC Press
Taylor & Francis Group
Boca Raton London New York

CRC Press is an imprint of the
Taylor & Francis Group, an **informa** business

CRC Press
Taylor & Francis Group
6000 Broken Sound Parkway NW, Suite 300
Boca Raton, FL 33487-2742

First issued in paperback 2019

© 1999 by Taylor & Francis Group, LLC
CRC Press is an imprint of Taylor & Francis Group, an Informa business

No claim to original U.S. Government works

ISBN-13: 978-1-56670-369-7 (hbk)
ISBN-13: 978-0-367-39990-0 (pbk)

Library of Congress Cataloging-in-Publication Data

Bregman, Jacob I.
    Environmental impact statements.—2nd ed. / Jacob I. Bregman.
        p. cm.
    Includes index.
    ISBN 1-56670-369-7
    1. Environmental impact statements—United States. I. Title.
TD194.55.B74 1999
333.7′14′0973—dc21
                                  98-51739
                                        CIP

Visit the CRC Press Web site at www.crcpress.com

Library of Congress Card Number 98-51739

**Visit the Taylor & Francis Web site at
http://www.taylorandfrancis.com**

**and the CRC Press Web site at
http://www.crcpress.com**

Author

Company president, university professor, politician, and grandfather, Jack Bregman, B.S., M.S., and Ph.D., is the epitome of the American success story. The only son of Russian immigrants, he is the author of over 60 publications, holds several patents, and has written 7 books. An internationally recognized expert in his field, Dr. Bregman is listed in several of the *Who's Who* publications, has been elected to membership in the honor societies of Sigma Xi and Phi Lambda Upsilon, and is a Fellow of the American Institute of Chemists.

After receiving two battle stars as a soldier in Europe during World War II, and being on a torpedoed troop ship in the English Channel, Dr. Bregman was awarded his Ph.D. in Chemistry by the Polytechnic Institute of Brooklyn, worked for several years in prominent research laboratories, and served in the Johnson Administration as Deputy Assistant Secretary for Water Quality. Since the mid-1960s, he has founded two highly successful environmental consulting firms, the second of which bears his name. He has been in government at the federal, state, and local levels, having won four elections and losing three.

Happily married for 50 years, Dr. Bregman and his wife, Mona, reside in Bethesda, MD, where their company is also located. When the question of retirement is raised, Dr. Bregman's standard comment is that perhaps when he reaches the age of 80, he may consider semiretiring to 40 hours per week in order to concentrate a little more on his writing and teaching.

Acknowledgment

Acknowledgment and appreciation for invaluable assistance are given to those who contributed directly or indirectly, currently or in the past, to the second edition of this book. Special thanks and appreciation are given to Mrs. Mona Bregman and Mrs. Dorothy Mackenthun for their support, encouragement, and tolerance of evenings and weekends devoted to its development. In addition, the questions and comments of students in the Graduate School of Engineering at George Washington University, who have taken the course in Environmental Impact Studies under Dr. Bregman, have been particularly valuable in shaping this text. Mr. Robert Edell and Ms. Michon Washington prepared much of the new material on environmental justice. Mr. Richard Mandel spent a considerable amount of effort in updating information for portions of this text. Ms. Samora Thompson and Ms. Linda Collins devoted long hours to its production. Mr. Ken Mackenthun assisted in the preparation of the first edition of this book.

List of Exhibits

Contents

Chapter 4

Chapter 5

Chapter 6

Appendix C

1 Purpose of the Environmental Impact Statement

1.1 THE CHANGE IN ATTITUDE

Until 1969, the national philosophy concerning negative environmental effects of major projects such as highways, industrial plants, shopping centers, housing developments, and so on, was to ignore them during the planning stages of the project. After the work was completed and the negative environmental effects were apparent, the attitude generally was one of "Too bad, but it couldn't be avoided." On a rare occasion, small mitigative measures would be installed after the fact.

Recognizing that this attitude of "build now and worry about it later" was almost universal, the U.S. Congress enacted the National Environmental Policy Act (NEPA) in 1969. This Act, in effect, said that environmental impacts should be considered *before* the plans for a project were finalized. "Look before you leap" became the approach to be taken.

1.2 WHAT IS NEPA?

Basically, the NEPA approach is a common sense approach. It requires that one think through the environmental consequences before taking action. If those consequences include undesirable effects, then NEPA requires that either consideration be given to mitigating measures that can be built into the action, or that alternatives to the action be considered that would produce similar end results but be less damaging to the environment. The national policy became one of answering the question "What is the impact on the environment of the planned action and how can it be minimized?" before starting a major action. This is a common sense approach that required a law to make it happen.

The document that examines the consequences of an action is called an Environmental Impact Statement (EIS). As a result of NEPA and the regulations by the Council on Environmental Quality (CEQ) that implement it, the EIS examines alternatives to the project as well as the planned action (termed the "Preferred Alternative" in the EIS). The alternatives are usually two or three in number and may include process changes, alternate geographical locations, change in size, and so on. In addition, the "Do Nothing" alternative must be examined, that is, what happens if the situation remains as is and no action is taken.

After a review of the information in the EIS concerning the various alternatives, an intelligent decision on how to proceed can be made. Most of the time, the Preferred Alternative is selected, but mitigating measures are applied to minimize negative environmental effects. But for the EIS, these mitigating measures might not have been included in the project. On occasion, one of the other alternatives is selected to replace the Preferred Alternative. In rare instances, the Do Nothing alternative prevails and no action is taken.

There are many occasions when there is a reasonable doubt about the possible need for the performance of an EIS. In those situations an Environmental Assessment (EA) may be performed first. The EA is reviewed by the lead agency. It determines whether there may be negative impacts on the environment, in which case an EIS is required. On the other hand, the EA may show no impacts worth considering. In that case, a Finding of No Significant Impact (FONSI) is documented, published in the *Federal Register*, and the project moves ahead.

1.3   ENVIRONMENTAL ASSESSMENTS

The Environmental Assessment covers the same basic topics as the Environmental Impact Statement. However, it does not require a scoping meeting, publication in the *Federal Register*, response to comments, and so on. Furthermore, the EA usually is limited to existing information rather than to the development of new data as is the case with the EIS. Generally, a draft EA is prepared and submitted to the agency within a six week to three month period. After agency review, the EA is finalized. By contrast, a draft EIS may require six months to one year or more to prepare. The additional time for various reviews and finalization may add another six months or more to the total elapsed time.

If the EA results in a preparation of a FONSI, the preparation can be completed in a week or so.

1.4   THE BENEFITS OF NEPA

NEPA has caused federal agencies to incorporate environmental values into their decision-making processes. For most agencies, the NEPA review is now an integral part of program planning. The primary benefit has been more protection of the environment in federal undertakings. This has come about because of the NEPA review process and the resultant changes in projects, such as alterations in project design, location, or operation; agency consideration of a greater range of alternatives; implementation of mitigation measures; and enhanced opportunity for public involvement in the decision-making process. An additional benefit has been a reduction of some project costs because of changes made in projects. The NEPA review process also has enabled agencies to address compliance with other environmental laws as part of a single review process rather than as separate reviews under each law, thereby reducing the amount of paperwork, staff time, and effort (EPA, 1989).

Almost anyone can think of actions taken before NEPA was instituted that would have been modified or changed if an EIS had been required. The following three actions are readily apparent:

1. The mounds of radioactive dirt that have almost destroyed Grand Junction, CO, as a residential and resort center would not have been put there if an EIS had been done on the uranium extraction project at that location.
2. Chemical plants found in the floodplains of the Ohio River would not have been placed there.
3. Thousands of acres of wetlands that were destroyed for highways and housing projects would have been saved.

In a similar manner, the following examples (all of which were in EIS performed by the author of this book) illustrate the value of an EIS or an EA:

- A highway that would have required destruction of a small, historic, black church was rerouted.
- When Union Station in Washington, D.C. was restored, the ceilings were kept in a manner faithful to the original David Burnham concept.
- A Congressional Award was presented to EPA Region 5 for the sponsorship of a study that developed alternative wastewater treatment systems for small resort communities on lakes in the Midwest. It was estimated that if these concepts were applied to similar locations across the country, the result would be a savings of billions of dollars.

1.5 EIS—WHEN AND WHO

When is an EIS type study required? NEPA and the implementing regulations call for one in the case of every major federal action that may have an impact on the environment. Federal actions are defined as both project and programmatic types. In addition, the issuance of federal permits to a non-federal entity, for example, the National Pollutant Discharge Elimination System (NPDES) permits triggers the EIS process.

Who is responsible for performing the EA or EIS? The answer is the government agency that is the most concerned with a particular action. Sometimes other agencies join it in the role of cooperating agencies. An industrial firm that is going to require a federal permit often may do the equivalent of an EA and submit it to the government.

1.6 OTHER JURISDICTIONAL EQUIVALENTS

Many cities and states have developed their own variations of NEPA. For example, some cities require potential developers to submit a document that describes the possible environmental effects of projects that will require city approval of one sort or another. States such as California and Maryland have their own versions of EIS for certain types of projects that may require state permits. The State of Maryland has issued a description of what goes into such an EA, which is very comprehensive and will be referred to in several sections of this book.

According to the U.S. Environmental Protection Agency (EPA), in 1989, 11 states had passed laws with comprehensive environmental review requirements (EPA, 1989). Limited environmental review requirements were established by executive order or other administrative directives in 14 other states.

The Comprehensive Environmental Response, Compensation, and Liability Act (CERCLA) is an example of a federal activity that has its own version of NEPA and EIS. This is done for two reasons:

1. The specialized complexity of activities under the act require NEPA-equivalency studies that are directly related to the specific Act.
2. Application of the NEPA–EIS process, as such, might create delays that would seriously impact projects being carried out under this Act.

For those reasons, the EPA has decided to include NEPA-equivalency studies in the performances of activities under the Act. Details will be presented later in Chapter 13.

1.7 EPA'S ROLE

The EPA, like other federal agencies, prepares and reviews NEPA documents. However, the EPA also takes a lead role in the NEPA review process for EIS from any agency. In addition, under Section 309 of the Clean Air Act as amended (1970 et seq.), the EPA is required to review and publicly comment on the environmental impacts of major federal actions including actions which are the subject of EIS. If the EPA determines that the action is environmentally unsatisfactory, it is required by Section 309 to refer that matter to the Council on Environmental Quality (CEQ). In the period between 1974 and 1989, there were 24 NEPA referrals to the CEQ, of which 14 were submitted by the EPA (EPA, 1989).

Also, in accordance with a Memorandum of Agreement between the EPA and the CEQ, the EPA carries out the operational duties associated with the administrative aspects of the EIS filing process. The Office of Federal Activities in the EPA has been designated as the official recipient in the EPA of all EISs prepared by federal agencies.

1.8 THE PUBLIC'S ROLE

The public has an important role in the NEPA process in providing input during the scoping process on what issues should be addressed in an EIS, and later commenting on the findings in an agency's NEPA documents. Later in this book, an entire chapter will be devoted to the subject of public participation. The public can participate further in the NEPA process by attending NEPA-related hearings or public meetings and by submitting comments directly to the lead agency. The lead agency must take into consideration all comments received from the public and other parties on NEPA documents during the comment period after publication of the notice in the *Federal Register* of the availability of the draft EIS. Comments must be included in the Appendix to the EIS. In addition, substantive comments from the public must be addressed in the EIS itself.

1.9   THE EIS IN FEDERAL DECISION MAKING

It should be noted that the EIS by itself is not the only determining factor as to whether or not a project goes forward, although it certainly is an influential factor.

The CEQ NEPA regulations require federal agencies to make environmental review documents and any comments and responses a part of the record in formal rule making and adjudicatory proceedings. These documents also must accompany the proposal through the federal agency's review process. In making its decision on a proposal, an agency must consider a full range of alternatives including those evaluated in the NEPA review (EPA, 1989).

Agencies also are encouraged to prepare broad EISs covering policy or programmatic actions and to tier subsequent NEPA reviews to individual actions included within the program or policy. For legislative proposals, the NEPA process is integrated with the legislative process of Congress.

REFERENCE

Facts about the National Policy Act, U.S. Environmental Protection Agency, 1989.

2 The Legal Basis for Environmental Impact Statements

In this chapter, the three federal documents that form the legal basis for the requirement for an Environmental Impact Statement (EIS) will be described:

1. The National Environmental Policy Act (NEPA).
2. The Council on Environmental Quality (CEQ) Regulations for Implementing NEPA.
3. Executive Order 11514 as amended by Executive Order 11991, entitled "Protection and Enhancement of Environmental Quality That Requires Federal Agencies to Implement NEPA."

Because of their importance to the subject of this book, the first two documents are reproduced in full in the appendices. The discussion of those three documents will be followed in this chapter by examples of agency regulations for their implementation. Finally, a brief discussion will present similar requirements by states and municipalities.

2.1 THE NATIONAL ENVIRONMENTAL POLICY ACT

A major turning point in the United States concern for the environment occurred in 1969 when Congress enacted the National Environmental Policy Act of 1969, otherwise known as NEPA. For the first time in our history, a law was enacted which required that environmental effects be considered before any activity requiring a federal permit could be undertaken. "Think before you act" became the national policy. A discussion of the key features of the Act follows.

NEPA established a nationwide policy for promoting environmental considerations for federal decision making. NEPA represents the national environmental goals and policies which are intended to mitigate mistakes of the past and to avoid possible problems in the future through thoughtful and coordinated planning efforts. The essence of the Act, and indeed of the entire environmental movement in this country, is stated in the Purpose of the Act (Section 2) as follows:

"To declare a national policy which will encourage productive and enjoyable harmony between man and his environment; to promote efforts which will prevent or eliminate damage to the environment and biosphere and stimulate the health and welfare of man; to enrich the understanding of the ecological systems and natural resources important to the Nation; and to establish a Council on Environmental Quality."

Under Title I of the Act, the definition of the national environmental policy is expanded as follows:

"The Congress, recognizing the profound impact of man's activity on the interrelations of all components of the natural environment, particularly the profound influences of population growth, high-density urbanization, industrial expansion, resource exploitation, and new and expanding technological advances and recognizing further the critical importance of restoring and maintaining environmental quality to the overall welfare and development of man, declares that it is the continuing policy of the Federal Government, in cooperation with State and local governments, and other concerned public and private organizations, to use all practicable means and measures including financial and technical assistance, in a manner calculated to foster and promote the general welfare, to create and maintain conditions under which man and nature can exist in productive harmony, and fulfill the social, economic, and other requirements of present and future generations of Americans.

(b) In order to carry out the policy set forth in this Act, it is the continuing responsibility of the Federal Government to use all practicable means, consistent with other essential considerations of national policy, to improve and coordinate Federal plans, functions, programs, and resources to the end that the Nation may—

(1) fulfill the responsibilities of each generation as trustee of the environment for succeeding generations;
(2) assure for all Americans safe, healthful, productive, and aesthetically and culturally pleasing surroundings;
(3) attain the widest range of beneficial uses of the environment without degradation, risk to health or safety, or other undesirable and unintended consequences;
(4) preserve important historic, cultural, and natural aspects of our national heritage, and maintain, whenever possible, an environment which supports diversity, and variety of individual choice;
(5) achieve a balance between population and resource use which will permit high standards of living and a wide sharing of life's amenities; and
(6) enhance the quality of renewable resources and approach the maximum attainable recycling of depletable resources.

(c) The Congress recognizes that each person should enjoy a healthful environment and that each person has a responsibility to contribute to the preservation and enhancement of the environment."

Section 102 of the Act then requires that "the policies, regulations, and public laws of the United States shall be interpreted and administered in accordance with the policies set forth in this Act." The section then goes on to implement this requirement by creating the environmental impact report process as follows:

"Include in every recommendation or report on proposals for legislation and other major Federal actions significantly affecting the quality of the human environment, a detailed statement by the responsible official on:

     (i) The environmental impact of the proposed action,
    (ii) Any adverse environmental effects which cannot be avoided should the pro-
        posal be implemented,
   (iii) Alternatives to the proposed action,
   (iv) The relationship between local short-term uses of man's environment and the
        maintenance and enhancement of long-term productivity, and
    (v) Any irreversible and irretrievable commitments of resources which would be
        involved in the proposed action should it be implemented."

The federal official responsible for an EIS is required to obtain the views of all federal, state and local agencies that may be affected. States are allowed to prepare their own EISs, provided that the responsible federal official is involved in the process and evaluates the statement.

Title II of NEPA establishes the Council on Environmental Quality (CEQ) in the Executive Office of the President. It is composed of three members, one of whom is the chairperson. The Council has the responsibility of preparing an annual Environmental Quality Report for transmittal by the President to the Congress. It also gathers information on trends in environmental quality, reviews federal programs for compliance with NEPA, conducts studies, and recommends national environmental policies and legislation. In exercising its powers, the CEQ consults with the Citizens Advisory Committee that was established by Executive Order No. 11472, on May 29, 1969.

The full text of the National Environmental Policy Act, current as of March 24, 1998, is found in Appendix 1 at the end of this book.

2.2   CEQ REGULATIONS FOR IMPLEMENTING NEPA

CEQ has issued regulations under the authority of NEPA and Executive Order 11514 as amended by Executive Order 11991. The CEQ Regulations apply to all federal agencies. Almost every federal agency has issued its own set of regulations that clarify how it will comply with the CEQ requirements. They all are similar in nature, follow the CEQ regulations quite closely and differ only in items that are unique to each agency.

The purpose of the CEQ regulations is to tell federal agencies what they must do to comply with NEPA. The regulations state that:

"NEPA procedures must insure that environmental information is available to public officials and citizens before decisions are made and before actions are taken. The information must be of high quality. Accurate scientific analysis, expert agency comments, and public scrutiny are essential to implementing NEPA. Most important, NEPA documents must concentrate on the issues that are truly significant to the action in question, rather than amassing needless detail. The NEPA process is intended to help public officials make decisions that are based on understanding of environmental consequences, and take actions that protect, restore, and enhance the environment. These regulations provide the direction to achieve this purpose."

A key provision of the regulations is that paperwork shall be kept to a minimum and the EISs are to be written in a manner that is understandable to the general public. Consequently, EISs generally are limited to 150 pages, have executive summaries, and are easy to understand. The detailed technical information that supports the EIS conclusions usually is found in an appendix to the EIS and has no page limit.

The CEQ regulations require that "Agencies shall integrate the NEPA process with other planning at the earliest possible time to insure that planning and decisions reflect environmental values, to avoid delays later in the process, and to head off potential conflicts." This timing is critical so that environmental factors may be considered while the opportunity exists to make adjustments in the project to accommodate environmental concerns.

The regulations allow agencies to perform an environmental assessment (EA) as a part of the process of determining whether an EIS should be prepared. The result of the EA is either the preparation of a Finding of No Significant Impact (FONSI) or else the preparation of an EIS. When it is obvious that an EIS will have to be prepared, the agency may go directly to the EIS step and skip the EA.

Each EIS has a lead agency which supervises its preparation. When more than one agency is involved in a project, a determination of the lead agency is made by the agencies involved. The remaining agencies then become cooperating agencies. In case of a disagreement as to who should be the lead agency, the determination is made by the CEQ.

Cooperating agencies include those that have jurisdiction over any phase of the project, or else have special expertise with respect to any environmental issue. They participate in the NEPA process and may develop some of the information required for the EIS. The U.S. Environmental Protection Agency (EPA), the U.S. Fish and Wildlife Service, and the Corps of Engineers are the most frequent cooperating agencies.

Scoping meetings are held very early in the EIS process for the purpose of determining the scope of the issues to be addressed and for identifying the significant environmental factors related to a proposed action. As soon as a decision is made by an agency to prepare an EIS, it must publish a Notice of Intent in the *Federal Register*. The scoping meeting is the next step in the EIS process. The lead agency invites "the participation of affected Federal, State, and local agencies, any affected Indian tribe, the proponent of the action, and other interested persons (including those who might not be in accord with the action on environmental grounds.)"

The scoping meeting is intended to accomplish the following objectives:

- "Determine the scope and the significant issues to be analyzed in depth in the environmental impact statement.
- Identify and eliminate from detailed study the issues which are not significant or which have been covered by prior environmental review.
- Allocate assignments for preparation of the environmental impact statement among the lead and cooperating agencies.
- Indicate any public environmental assessments and other environmental impact statements which are being or will be prepared that are related to but are not part of the scope of the impact statement under consideration.
- Identify other environmental review and consultation requirements so the lead and cooperating agencies may prepare other required analyses and studies concurrently with, and integrated with, the environmental impact statement.

- Indicate the relationship between the timing of the preparation of environmental analyses and the agency's tentative planning and decision-making schedule."

The CEQ regulations then define an EIS and discuss its implementation. EISs are required for major federal actions. In addition to projects requiring federal permits, this may include the adoption of new agency programs or regulations. Timing is critical so that the EIS will be able to affect the project. The regulations state that "An agency shall commence preparation of an environmental impact statement as close as possible to the time the agency is developing or is presented with a proposal so that preparation can be completed in time for the final statement to be included in any recommendation or report on the proposal. The statement shall be prepared early enough so that it can serve practically as an important contribution to the decision-making process and will not be used to rationalize or justify decisions already made."

EISs are to be written in plain language so everyone understands them. They are to proceed in two stages as follows:

1. Draft EIS according to the decisions made in the scoping process.
2. Final EIS should respond to comments made on the draft EIS.

In addition, supplementary EISs may be prepared if the situation warrants it. The proposed format includes the following:

1. Cover sheet.
2. Summary.
3. Contents.
4. Purpose of and need for action.
5. Alternatives including proposed action.
6. Affected environment.
7. Environmental consequences.
8. List of preparers.
9. List of agencies, organizations, and persons to whom copies of the statement are sent.
10. Index.
11. Appendices (if any).

Each item in the preceding list then is defined in detail in the regulations.

The cover sheet is to contain a list of the responsible agencies; the title of the proposed action and its location; identification of the person who can supply more information; a designation of the EIS as draft, final or supplementary; an abstract; and the date by which comments on the EIS must be received.

The summary contains the major conclusions, areas of controversy, and issues to be resolved. It is not to exceed 15 pages.

The discussion of alternatives must present "the environmental impacts of the proposal and the alternatives in comparative form, thus sharply defining the issues and providing a clear basis for choice among options by the decision-maker and the public." If preferred alternatives exist, they are to be identified. Also, appropriate mitigative actions for negative impacts are to be shown.

The affected environment means the environment of the area that may be affected as it exists prior to the proposed action.

The environmental consequences of each of the alternatives form the basis for their comparison. Especially critical factors should include any adverse environmental effects which cannot be avoided, the relationship between short-term uses of man's environment and the maintenance and enhancement of long-term productivity, and any irreversible or irretrievable commitments of resources which would be involved in the proposal should it be implemented. Both direct and indirect effects are to be considered. Conflicts with plans, policies, and controls for the area concerned are to be noted.

Consequences are defined so as to include environmental effects, energy and resource requirements, as well as urban quality, historic, and cultural resources.

The list of preparers should include a description of their qualifications. The appendix has all of the back-up material prepared or obtained during the EIS and is to be circulated with the EIS.

Circulation of the draft and final EISs are to be made according to the following list:

> "(a) Any Federal agency which has jurisdiction by law or special expertise with respect to any environmental impact involved and any appropriate Federal, State or local agency authorized to develop and enforce environmental standards.
> (b) The applicant, if any.
> (c) Any person, organization, or agency requesting the entire environmental impact statement.
> (d) In the case of a final environmental impact statement, any person, organization, or agency which submitted substantive comments on the draft."

Whenever a broad EIS has been prepared (such as a program or policy statement) and a subsequent statement or EA is then prepared on an action included within the entire program or policy (such as a site-specific action), the subsequent statement or EA need only summarize the issues discussed in the broader statement and incorporate discussions from the broader statement by reference. It should only concentrate on the issues specific to the subsequent action. Similarly, in order to reduce EIS size, material which is readily available to the public may be incorporated by reference.

Cost–benefit analyses details may be placed in the appendix; only the results need to be discussed in the EIS. Methodologies used in the EIS are to be named in the EIS and discussed in the appendix.

The regulations emphasize the need to prepare draft EISs concurrently with and integrated with the related documents required by the following:

- Fish and Wildlife Coordination Act.
- National Historic Preservation Act.

- Endangered Species Act.
- Other environmental review laws and executive orders.

The draft EIS must list all federal permits, licenses, and so on that must be obtained in order to implement the proposal.

The CEQ regulations contain very specific requirements relative to comments on the EIS:

- Inviting them.
- Duty of agencies to respond.
- Specificity of comments.
- Responses to them.

After preparation of the draft EIS, the agency sponsoring it must obtain comments from the following:

- Federal agencies.
- State and local agencies.
- Indian tribes, when affected.
- Any other interested agency.
- The applicant, if any.
- The interested public, by soliciting comments from persons and organizations.

Federal agencies are duty bound to comment on the draft EIS or to respond by saying that they have no comment. Comments are to be as specific as possible with regard to methodology, the need for additional information, and possible mitigation measures necessary to allow permits or licenses to be issued.

Comments to the draft EIS are to be responded to by any of the following methods:

- Modification of alternatives, including the proposed action.
- Development of new alternatives.
- Modification of analyses in the EIS.
- Factual corrections.
- Explaining why the comments do not apply.

The comments and the responses should be incorporated into the final EIS. This is usually done in the appendix for those comments that are not accepted.

There may be cases where proposed major federal actions may cause unsatisfactory environmental effects, as determined by the EPA in its authority under the Clean Air Act as amended, (1970 et seq.). In that event, or if another federal agency makes a similar determination in its NEPA review, and differences cannot be resolved with the lead agency, the matter is referred to the CEQ for judgment. Every possible attempt is to be made to minimize this from happening, with emphasis on mitigating the unfavorable environmental consequences or else switching to a more favorable alternative.

If referral does occur, the CEQ must start action within 25 days to resolve the matter and complete that action in 60 days.

Part 1505 of the CEQ regulations sets forth the requirements that the NEPA process be included in all of each federal agency's principal programs likely to have a significant effect on the human environment. The relevant environmental factors are to be considered in deciding between alternatives. Public input is encouraged.

A public record of decision (ROD) must be made in cases where EISs were required. The ROD must discuss the alternatives and describe any practicable means of avoiding or minimizing environmental harm, including possible monitoring and enforcement programs. Permits and funding of the actions are to have mitigation and monitoring as conditions of approval where necessary.

Part 1506 of the regulations covers other requirements of NEPA. The first requirement places limitations on any actions that may be taken on a proposal subject to NEPA until the EIS process is completed, thus preventing a fait accompli and ensuring that the NEPA process will work. Another requirement encourages federal agencies to cooperate with state and local agencies to reduce duplication between NEPA and comparable state and local requirements.

Agencies are allowed to adopt EISs prepared by other agencies if the proposed actions are essentially the same, or to combine them with other agency documents. EISs are to be prepared by contractors chosen by the lead agency or, where appropriate, by a cooperating agency. In any event, there is to be no conflict of interest concerning the contractor.

Public involvement in the NEPA procedures is stressed. Adequate public notices of the availability of the NEPA documents is emphasized. Actions with national concern effects are to have a notice published in the *Federal Register.* In addition, national organizations that may be interested are to be notified. For actions of local interest, notices are to given to Indian Tribes of effects that may occur on reservations. In addition, notices are to be published in local newspapers (rather than legal papers) and also are to be given to community organizations, small business associations, newsletters, and the like. Furthermore, information about the EIS may be distributed by direct mailing to owners and occupants of nearby affected properties. A notice may be posted at the location where the action will take place.

The federal agency may decide to hold public hearings if substantial environmental controversy exists, if there is substantial interest in a hearing, or if another agency with jurisdiction over the action feels that a hearing will be helpful. When a public hearing is held on an EIS, a notice to the public must be given at least 15 days in advance.

EISs, comments received, and all of the underlying information are to be made available to the public either without charge, or for the actual costs of reproduction. This may be done under the Freedom of Information Act.

From time to time, the CEQ may provide federal agencies with further guidance concerning NEPA by using any of several procedures available to it.

In certain circumstances, proposed congressional legislation may require an EIS or its equivalent, which must be available in time for congressional hearings on the legislation. No scoping meetings are required. Further, the document is called a detailed statement instead of a draft EIS. However, conventional draft and final EISs are to be prepared under any of the following conditions:

> "(i) A Congressional Committee with jurisdiction over the proposal has a rule requiring both draft and final environmental impact statements.
>
> (ii) The proposal results from a study process required by statute (such as those required by the Wild and Scenic Rivers Act and the Wilderness Act.)
>
> (iii) Legislative approval is sought for Federal or Federally assisted construction or other projects which the agency recommends be located at specific geographic locations. For proposals requiring an environmental impact statement for the acquisition of space by the General Services Administration, a draft statement shall accompany the Prospectus or the 11(b) Report of Building Project Surveys to the Congress, and a final statement shall be completed before site acquisition.
>
> (iv) The agency decides to prepare draft and final statements. Comments on the legislative statement are given to the lead agency which forwards them along with its own responses to the Congressional committees with jurisdiction."

EISs along with comments and responses are filed with the EPA, which delivers copies to the CEQ.

The EPA publishes a notice each week that lists the EISs filed during the preceding week. Decisions by agencies on the proposed actions cannot be made until 90 days after a draft EIS or 30 days after a final EIS. Exceptions are made in the case of appeals by other agencies or the public. Exceptions also may be made when rule making is for protection of public health or safety. In any event, not less than 45 days is to be allowed for comments on draft statements. The lead agency may extend the prescribed periods. The EPA may reduce the periods for compelling reasons of national policy. Provision is made for emergency situations where it is necessary to take an action with significant environmental impacts without observing the regulations.

Part 1507 of the regulations requires all agencies of the federal government to comply with the CEQ regulations. Each agency has a degree of flexibility in adapting its implementation procedures to NEPA.

A few of the definitions in Part 1508 are worthy of repetition here as they are particularly important in the EIS process. They are as follows:

> "Categorical Exclusion" means "a category of actions which do not individually or cumulatively have a significant effect on the human environment and which have been found to have no such effect in procedures adopted by a Federal agency in implementation of these regulations (§1507.3) and for which, therefore, neither an environmental assessment nor an environmental impact statement is required."

> "Finding of No Significant Impact" means "a document by a Federal agency briefly presenting the reasons why an action, not otherwise excluded, will not have a significant effect on the human environment and for which an environmental impact statement therefore will not be prepared. It shall include the environmental assessment or a summary of it and shall note any other environmental documents related to it."

"Major Federal Actions" are defined as including the following categories: "

"(1) Adoption of official policy, such as rules, regulations, and interpretations; treaties and international conventions or agreements; and formal documents establishing an agency's policies which will result in or substantially alter agency programs.

(2) Adoption of formal plans, such as official documents which guide or prescribe alternative uses of Federal resources, upon which future agency actions will be based.

(3) Adoption of programs, such as a group of concerted actions to implement a specific policy or plan.

(4) Approval of specific projects, such as construction or management activities located in a defined geographic area. Projects include actions approved by permit or other regulatory decision as well as Federal and Federally assisted activities."

"Mitigation" includes:

"(a) Avoiding the impact altogether by not taking a certain action or parts of an action.

(b) Minimizing impacts by limiting the degree or magnitude of the action and its implementation.

(c) Rectifying the impact by repairing, rehabilitating, or restoring the affected environment.

(d) Reducing or eliminating the impact over time by preservation and maintenance operations during the life of the action.

(e) Compensating for the impact by replacing or providing substitute resources or environments."

"Notice of Intent" means a notice that an environmental impact statement will be prepared and considered. The notice shall briefly:

"(a) Describe the proposed action and possible alternatives.

(b) Describe the agency's proposed scoping process including whether, when, and where any scoping meeting will be held.

(c) State the name and address of a person within the agency who can answer questions about the proposed action and the environmental impact statement."

The full text of the CEQ regulations on implementing NEPA may be found in Appendix B at the end of this book.

2.3   EXECUTIVE ORDER 11514, PROTECTION AND ENHANCEMENT OF ENVIRONMENTAL QUALITY

Executive Order 11514 was promulgated on March 5, 1970 and was amended by Executive Order 11991 [Sections 2(g) and 3(h)] on May 24, 1977. It requires federal agencies to conform with NEPA under the guidance of the CEQ. Details of the Executive Order follow.

Section 1 sets forth what the federal government policy of environment is to be as follows:

"The Federal Government shall provide leadership in protecting and enhancing the quality of the Nation's environment to sustain and enrich human life. Federal agencies shall initiate measures needed to direct their policies, plans and programs so as to meet national environmental goals."

Section 2 makes it the responsibility of all federal agencies to monitor, evaluate, and control on a continuing basis their agencies' activities so as to protect and enhance the quality of the environment. Agencies are to develop programs and measures to protect and enhance environmental quality and shall assess progress in meeting the specific objectives of such activities.

Agencies also are to develop procedures to ensure the provision of timely public information concerning federal plans and programs with environmental impacts in order to obtain the views of interested parties. Procedures are to include, whenever appropriate, provision for public hearings. Federal agencies are to encourage state and local agencies to adopt similar procedures for informing the public concerning their activities affecting the quality of the environment.

Section 3 of the Executive Order sets forth the responsibilities of the CEQ. It gives the CEQ overview responsibility for federal policies and activities directed toward pollution control and environmental quality. With regard to environmental impacts, the CEQ is to:

"(e) Promote the development and use of indices and monitoring systems (1) to assess environmental conditions and trends, (2) to predict the environmental impact of proposed public and private actions, and (3) to determine the effectiveness of programs for protecting and enhancing environmental quality.

(f) Coordinate Federal programs related to environmental quality.

(h) Issue regulations to Federal agencies for the implementation of the procedural provisions of the Act."

2.4 NEPA REGULATIONS BY OTHER AGENCIES

As indicated above, the CEQ regulations call for the development of regulations by each federal agency on how it will implement the NEPA process. This has been done, but there is a substantial degree of variation from agency to agency in regard to the type of projects and policy matters that will have to conform to the EIS process. The individual agency regulations, therefore, make provision for these variations while, at the same time, adhering to NEPA and the CEQ regulations.

A listing of federal agency regulations implementing NEPA is located in Appendix 3 at the end of this book.

The organization that is the most involved with NEPA is the EPA. A discussion follows that describes the approach the EPA uses to develop regulations for the implementation of the NEPA process.

EPA's compliance procedures with NEPA are contained in 40 CFR 6. These procedures establish a straightforward, step-by-step approach for ensuring that agency decision making includes careful consideration of all environmental effects of proposed actions, analysis of potential environmental effects of proposed actions and their alternatives, provision for public understanding and scrutiny, and avoidance or minimization of adverse effects to the extent possible. The process is designed to incorporate consideration of environmental factors into the decision-making process at the earliest possible point.

For all of the EPA programs addressed under 40 CFR 6, the environmental review process includes the identification and exercise of decision-making authority by a "responsible official." This person may be an EPA Regional or Headquarters official, supplemented by staff, and assisted by other agency staff or consultants.

The first document prepared in the NEPA process is an "Environmental Information Document" which is prepared by applicants, grantees, or permit applicants and submitted to the EPA. The environmental information document must include adequate information to enable the responsible official to prepare the environmental assessment. The environmental information document, at a minimum, should include the following information:

- Overview of the proposed action, including purpose and need.
- Description of the existing environment.
- Description of the future environment.
- Development and evaluation of alternatives.
- Description of the environmental impacts of the action.

Key requirements for the environmental information document include ensuring a thorough evaluation, particular consideration of indirect impacts, and evaluation of the no-action alternative. If the proposed undertaking receives a categorical exclusion, then an environmental information document need not be prepared.

Based on the information prepared by the applicant, grantee, or permittee, the EPA then carries out an environmental review, and prepares an EA of the proposed action. The EA is a document that is made available to the public as a record of the EPA decision-making process carried out in review of the environmental information document. Based on this review, a decision is made as to whether an EIS or FONSI is required. If it can be determined ahead of time that an EIS will be required, it is not necessary to prepare a formal EA. When deciding whether an EIS is required for a specific undertaking, the EPA considers a number of issues. In general, an EIS is required in cases where significant, unavoidable adverse impacts are anticipated, and it is not considered feasible to mitigate these impacts.

Where it is determined that an EIS is necessary, the EPA must issue a Notice of Intent in the *Federal Register*. Following this announcement, the EIS is prepared either by EPA staff, with or without contractor assistance, or by a third party under an agreement with the applicant. Where the requirement for an EIS is determined early enough in the planning of the proposed project, it is possible to carry out a "piggyback" EIS, where the EIS is prepared jointly with the environmental information document.

Regardless of the particular approach taken, the key steps in the EIS process include the following:

- *Scoping,* an early, open process for determining the scope of issues to be addressed and for identifying the significant issues related to the proposed action.
- *Purpose and need* for the project.
- *Alternatives,* including the proposed action.
  - Alternatives considered by the applicant.
  - The no-action alternative.
  - Alternatives available to the EPA.
  - Alternatives available to other permitting agencies.
  - Identification of the *preferred alternative.*
  - Description of the *affected environmental* and *environmental consequences* of each alternative.
  - *Coordination* with other federal, state, and local agencies.
  - *Participation of the public* through hearings, meetings and other activities.

After completion of the EIS, the responsible official then prepares a concise public *Record of Decision* (ROD). The ROD includes mitigation measures implemented to make the selected alternative environmentally acceptable.

The final step in the general series of actions taken to comply with NEPA is *monitoring.* This includes all actions taken by the responsible official to ensure that decisions based on the EIS are properly implemented.

There are a series of EPA requirements for specific project and program EISs. Some of these will be discussed in later chapters in this book under topics concerning various portions of the EIS.

2.5   STATE AND MUNICIPAL EQUIVALENTS OF NEPA

Many states and municipalities have developed their own modifications of EIS requirements and utilize them as a part of the permit-granting process for major new construction. This includes facilities such as industrial plants, schools, highways, shopping centers, and so on. The requirements tend to vary from state to state and from municipality to municipality as a function of the environmental concerns that are of greatest importance to the local authorities.

As would be expected because of its pollution problems, local control is the most intensive in the state of California. Requirements have existed for many years for documents that are the equivalent of environmental assessments for most new construction, even of relatively small size. In a number of other states, delegation of the NEPA process by the EPA is an accomplished fact. The EPA maintains oversight authority on these environmental documents and may overrule the state if it believes that the documents are inadequate.

In connection with state activities, a brief discussion follows concerning the relationship between the EPA and a state when the state has been delegated NEPA responsibility. On the surface, the relationship is a straightforward one. The state has

NEPA responsibilities and the EPA basically has an oversight responsibility that allows it to overrule the state in a situation where such an action would be necessary because of the state's failure to comply properly with NEPA. The EPA also has the final EIS determination.

It is generally assumed that the state will have a more relaxed attitude towards NEPA than will the EPA, because of the state's proximity to the projects and the economic benefits deriving therefrom. This turns out to be the case in some situations but not in others. In 1980–1981, a team of EIS-skilled specialists under the overall direction of the author of this book undertook a study for the Regional Administrator of Region IV. The purpose was to assess the effectiveness of the NEPA environmental review process by the states of North Carolina, South Carolina, Georgia, Florida, Alabama, Mississippi, Kentucky, and Tennessee, some of which were delegated and others were not. The study was accomplished by evaluating the existing environmental review procedures that had been used on selected projects that were subject to the EPA NEPA compliance program and others that have not been subject to NEPA review.

Major issues examined included the following:

- Legislative and regulatory authority established by the states and federal government for 24 different environmental areas addressed under NEPA.
- Regulatory procedures required under state and federal NPDES programs and regulatory review of associated environmental resource areas.
- Level of protection and effectiveness of regulations established under state and federal review programs for 24 different environmental areas.
- Time and resource expenditures for NPDES and associated environmental reviews.
- Coordination of environmental review programs among agencies.
- Scope of mitigative authority to protect 24 different environmental areas.

The result of the study was a finding that the success of each state program was a function of the attitude of the state involved. This depends upon the relative

1. Environmental quality.
2. Environmental sophistication.
3. Comprehensiveness.
4. Budget of a state's program.

Some of the delegated states performed better environmental reviews than were done by the EPA in nondelegated states. Others did not do as well.

3 The Process of Preparing an EIS

This process can be a lengthy and expensive one. Recognizing this drawback, provisions have been made by most federal agencies to speed up the NEPA process by limiting the preparation of an EIS to those cases where negative impacts are likely to occur.

In the first part of this chapter, we will briefly describe the pre-EIS process that serves to eliminate many routine federal actions from the need for an EIS. That discussion will be followed by a discussion of the steps that should be taken in preparing an EIS.

3.1 PRE-EIS SCREENING

There are a number of different approaches taken by federal agencies to making a decision as to whether or not an EIS should be performed. The two that are obvious are:

1. Exemption by law, for example, for security purposes.
2. Emergencies, for example, national defense actions such as Desert Storm.

The most common other type is discussed below.

3.1.1 CATEGORICAL EXCLUSION (CX)

Many federal actions are of a routine nature and generate no significant impacts on the environment. Thus, moving an office from one floor to another in the same building would be an example of this type of action. To illustrate this, we present below a list of 29 such actions as developed by the U.S. Department of the Army (DA) based on the following criteria:

- Minimal or no individual or cumulative effect on environmental quality.
- No environmentally controversial change to existing environmental conditions.
- Similar to actions previously examined and found to meet the preceding criteria.

The DA list of categorical exclusions (CX) as cited in Appendix A of AR 200-2 (as of 1996) includes the following:

1. Normal personnel, fiscal, and administrative activities involving military and civilian personnel.

2. Law and order activities performed by military policy and physical plant protection and security personnel.
3. Recreation and welfare activities not involving off-road recreational vehicle management.
4. Commissary and Post Exchange (PX) operations except where hazardous material is stored or disposed.
5. Routine repair and maintenance of buildings, roads, airfields, grounds, equipment, and other facilities, except when requiring application or disposal of hazardous or contaminated materials.
6. Routine procurement of goods and services.
7. Construction that does not significantly alter land use.
8. Simulated war games and other tactical and logistical exercises without troops.
9. Training entirely of an administrative or classroom nature.
10. Storage of materials, other than ammunition, explosives, pyrotechnics, nuclear, and other hazardous or toxic materials.
11. Operations conducted by established laboratories within enclosed facilities where:
    a. All airborne emissions, waterborne effluents, external radiation levels, outdoor noise, and solid bulk waste disposal practices are in compliance with existing federal, state, and local laws and regulations.
    b. No animals that must be captured from the wild are used as research subjects.
12. Developmental and operational testing on a military installation, where the tests are conducted in conjunction with normal military training or maintenance activities.
13. Routine movement of personnel; routine handling and distribution of nonhazardous and hazardous materials in conformance with the DA, EPA, Department of Transportation, and state regulations.
14. Reduction and realignment of civilian and/or military personnel that fall below the thresholds for reportable actions as prescribed by statute or AR-510.
15. Conversion of commercial activities (CA) to contract performance of services.
16. Preparation of regulations, procedures, manuals, and other guidance documents that implement, without substantive change, the applicable DA or other federal agency regulations, procedures, manuals, and other guidance documents that have been environmentally evaluated.
17. Acquisition, installation, and operation of utility and communications systems, data processing, cable, and similar electronic equipment that use existing rights of way, easements, distribution systems, and facilities.
18. Activities that grant permits to identify the state of the existing environment without alteration of that environment or capture of wild animals.
19. Deployment of military units on a temporary duty (TDY) basis where existing facilities are used and the activities to be performed have no significant impact on the environment.

20. Grants of easements for the use of existing rights-of-way by vehicles; electrical, telephone, and other transmission and communication lines; transmitter and relay facilities; water, wastewater, stormwater, and irrigation pipelines, pumping stations and facilities; and for similar public utility and transportation uses.
21. Grants of leases, licenses, and permits to use existing Army controlled property for non-Army activities, provided there is an existing land-use plan that has been environmentally assessed and the activity will be consistent with that plan.
22. Grants of consent agreements to use a government-owned easement in a manner consistent with the existing Army use of the easement.
23. Grants of licenses for the operation of telephone, gas, water, electricity, community television antenna, and other distribution systems normally considered as public utilities.
24. Transfer of real property administrative control within the Army, or to another military department or other federal agency.
25. Disposal of uncontaminated buildings and other improvements for removal off-site.
26. Studies that involve no commitment of resources other than manpower.
27. Study and test activities within the procurement program for Military Adaptation of Commercial Items for items manufactured in the United States.
28. Development of table organization and equipment documents, no fixed location or size.
29. Grants of leases, licenses, and permits to use DA property for or by another governmental entity when such permission is predicated upon compliance with NEPA.

Of these 29 CXs, the following require a record of environmental consideration (REC), which is another DA safeguard for the NEPA process:

A-7, A-11(b), A-12, A-14, A-19, A-20, A-21 through A-29.

A REC describes the proposed action and anticipated time frame, identifies the proponent, and explains why further environmental analysis and documentation is not required. It is a signed statement to be submitted with project documentation.

3.2 ENVIRONMENTAL INFORMATION DOCUMENT (EID)

The EID was described in detail earlier in Chapter 2. It is prepared by applicants, grantees, or permits and submitted to the EPA. Other agencies have similar requirements for possible NEPA activities that originate from private sources, with other names given to these documents. Regardless of what the documents are called, they serve the purpose of giving the agency all of the project environmental information that already has been assembled by the applicant. The federal agency involved is placed in a much better position to determine whether to go to a categorical exclusion, an environmental assessment or an EIS.

3.3   ENVIRONMENTAL ASSESSMENT (EA)

The composition of EAs varies from agency to agency. They have the same basic contents as EISs, but they are performed in a much shorter time period, for example, six weeks to three months. This is because the EA is the forerunner of the EIS and determines whether an EIS will be required.

Major differences between EAs and EISs, in addition to the shortened time period, include the following:

- No *Federal Register* notices, scoping meetings, or public meetings are required. Some agencies do those anyhow.
- Almost all the data collected is that which is already available, rather than new material.
- No publication of the availability of the draft EA must be placed in the *Federal Register*.

The EA concludes with a recommendation that one of the two following courses be pursued:

1. Because of possible significant impacts, an EIS should be performed.
2. Because there will not be any significant impact, a FONSI should be prepared.

This allows the project to proceed without further environmental studies. It does not preclude the project proponent from being responsible for adopting mitigating measures to prevent unforeseen negative impacts.

The FONSI usually is widely distributed and may be subject to written changes, or even litigation. In the case of a FONSI, it is worthwhile to wait a reasonable period of time before proceeding with the project, so that any disagreement over the FONSI may be resolved.

3.4   THE ENVIRONMENTAL IMPACT STATEMENT PROCESS

While each federal agency has its own approach to an EIS process, agencies tend to differ only in small details. The generally accepted process is discussed below. The approach that will be taken is that of the person or firm that would do an EIS. A series of tasks will be presented and each one discussed in terms of content and approximate elapsed time for the task.

The major steps and analyses in the EIS process may be carried out in a series of discrete tasks as described next.

3.4.1   TASK 1   INITIAL MEETING WITH FEDERAL
AGENCY (CLIENT)

Within a day or two after an EIS project starts, a meeting should be held between the EIS project personnel and the federal agency (client) sponsoring the EIS. The purpose of the meeting will be to allow the client to transmit to the EIS preparer all available information on the project scope, existing site conditions, known feasible alternatives, and various studies and reports relevant to the project, including any other EIS near the site. Project issues will be defined and scoping meetings planned.

3.4.2   TASK 2   METHODOLOGY APPROVAL

Using the information developed in Task 1, the EIS methodology should be submitted to the client for approval within 15 days after project starts. Any necessary revisions required by the client are made then.

3.4.3   TASK 3   SCOPING MEETINGS

Prior to the scoping meetings, it is often useful to prepare a 5 to 10 page preliminary environmental analysis (PEA) that identifies the geographic area of the proposed project, reviews the alternatives, describes the important characteristics of the area, and discusses the significant project-related issues. It goes on to present a proposed outline of the EIS and gives a brief discussion of each item therein. The PEA then serves as the handout at the scoping meetings and the starting point for consideration of changes or additions to the EIS. Individuals and firms are identified and invited to participate in the scoping meeting. Arrangements are made for the meeting areas and newspaper advertisements. An agenda is prepared. The EIS scoping meeting(s) usually are held within 30 days after project starts. The specific times and places of the scoping meetings usually are in the areas to be served by the project. The scoping meetings are held in accordance with the CEQ's EIS scoping requirements (40 CFR 1501.7) and minutes or transcripts of the meeting are taken.

The purpose of the scoping meeting(s) is to determine the scope of the draft environmental impact statement (DEIS) and to identify the major project-related issues to be addressed and emphasized in it. Comments to this effect by the attending agencies and public are solicited. Invited agency representatives consist of all of the federal agencies that may have an interest in project impacts and/or participate in the EIS as cooperating agencies. State and local agencies invited include all of the pollution prevention, natural resource, historical and archaeological agencies, and any others who express an interest. Public participation includes groups or individuals.

The product of the scoping meeting is a brief paper (scoping report) that summarizes the significant alternatives and related issues. The paper reflects issues and the extent of coverage to be contained in the draft EIS. A brief discussion of the scoping process and the comments received from the public are included in the report.

3.4.4   Task 4   Data Collection and Description of the Existing Environment

The suitability of environmental data obtained from the client in Task 1 is evaluated for use in the description of existing environmental conditions, the evaluation of project alternatives, and the assessment of the alternatives' impacts upon the environment. Work in Task 4 focuses on the gathering of additional data, where necessary, in order to address the significant issues of the EIS. Every attempt should be made to avoid duplication of others' work in the gathering and analysis of data. A discussion of the data to be gathered and analyzed follows. It will be broken down into two major categories, identified as the natural environment and the person-made environment. The topics that go into each category are listed. They are discussed in detail in other chapters in this book.

3.4.4.1   Natural Environment

The natural environment as defined here consists of the geology and the biology of the project area.

3.4.4.2   Person-Made Environment

The discussion of the existing person-made environment includes, but is not limited to, the following topics: water quality (including surface waters and groundwater), noise, air quality, land use, historic preservation and archaeology, demography, housing, local economy and other socioeconomic aspects, hazards and nuisances, aesthetics and urban design, community services, and transportation.

3.4.5   Task 5   Assessment of Potential Environmental Impacts

The potential impacts of each proposed project alternative are assessed as well as those of the "do nothing" alternative, on each of the environmental and social components delineated above under Task 4. The level of impact analysis is site specific. Emphasis is placed on the key issues identified and discussed during the scoping phase of the project.

Identification is made of the potential short- and long-term impacts associated with the project. Short-term impacts resulting from the proposed project may be those associated with the construction phase, including such disturbances as noise, dust, erosion, and wildlife displacement. Long-term post-construction impacts may include such factors as pollution from stormwater runoff, air and surface water pollution, noise, consumption of energy, depletion and contamination of groundwater sources, overloading of roadways and other infrastructure, and placing of heavy demands on community services such as sewage treatment and the disposal of solid wastes. These impacts are further characterized as avoidable, unavoidable, and capable of being mitigated.

In doing the assessments, one should follow standards and utilize analytical procedures established and approved by the EPA and other federal agencies having applicable legal jurisdiction. The impacts on public services may be quantified by calculating usage based on actual per household demand data for these areas. Air quality and noise impacts are assessed in accordance with state requirements. A summary of the impacts of each alternative should be presented in the EIS.

Irreversible and irretrievable resource commitments resulting from the implementation of the proposed action are described. The consumption of resources is categorized in terms of environmental and human effects.

Mitigative actions should be included in the discussion. Where possible, their costs and benefits should be quantified. These facts will be critical in the final decisions as to which alternative to select.

3.4.6 TASK 6 PREPARATION OF THE PRELIMINARY DRAFT ENVIRONMENTAL IMPACT STATEMENT (PDEIS)

A preliminary draft EIS (PDEIS) is prepared that addresses the major issues concerning the proposed project at the alternative sites and for the no action alternative. The preliminary draft EIS format should follow CEQ rules and regulations for the preparation of EIS by federal agencies. The preliminary EIS also should be consistent with the procedures and requirements of the sponsoring agency. The format of the PDEIS usually will include the following items:

- Cover sheet.
- Summary.
- Table of contents.
- Purposes and need for action.
- Alternatives, including the proposed action.
- Affected environment.
- Environmental impacts, including a discussion of impacts that can and cannot be mitigated.
- List of preparers.
- List of agencies, organizations, and persons to whom copies of the EIS are sent.
- Index.
- Appendices.

The cover sheet identifies the person in the agency to whom comments on the DEIS are to be sent. The appendices are bound separately from the rest of the EIS and contain materials such as the following:

- Scoping meeting details.
- Detailed data from which information in the EIS is drawn.
- Letters from agencies and the public.

3.4.7   TASK 7   PREPARATION OF THE DRAFT EIS (DEIS)

After the sponsoring agency has reviewed the PDEIS and suggested changes, a final version is prepared and is labeled as the draft EIS (DEIS). After agency review and approval, copies are transmitted to interested parties. In addition, a notice is placed in the *Federal Register* that identifies the EIS, the agency, and the manner in which copies may be obtained. A date is given for the receipt of comments on the draft, usually 45 days after the issuance of the DEIS.

3.4.8   TASK 8   RESPONSE TO COMMENTS

The preparation of the preliminary final EIS commences with transmittal of comments received by the agency during the draft EIS public comment period. Administrative or policy questions are answered by the agency and given to the EIS preparer, who develops answers to the technical comments, which then are submitted to the agency for approval.

3.4.9   TASK 9   PRELIMINARY FINAL EIS (PFEIS)

A preliminary final EIS (PFEIS) includes the responses to comments, as well as presenting the full text of all comments in the appendix, is prepared. It is reviewed and approved by the agency.

3.4.10   TASK 10   PREPARATION OF THE FINAL EIS (FEIS)

The agency's approved revisions of the PFEIS are incorporated into the FEIS. Upon approval by the agency, the FEIS is distributed to the appropriate parties.

3.4.11   TASK 11   RECORD OF DECISION (ROD)

The agency prepares a ROD after the EIS is completed.

4 Public Participation

4.1 INTRODUCTION

An effective public participation program in an environmental impact statement (EIS) enhances the probability of a technically accurate, economically feasible, and socially and politically acceptable EIS. Public involvement in an EIS is very desirable for two reasons:

1. It makes the public a partner to the process. Rumors are laid to rest and the public has the actual facts about the proposed project. Inevitably, this decreases public tensions and hostility to the project.
2. The public often has good suggestions for items to be incorporated or stressed in the EIS. The result is a better product.

The National Environmental Policy Act of 1969 (NEPA) mandates public involvement in assessing the environmental consequences of major federal actions. Consequently, public review and input on environmental reviews has become an integral part of the evaluation process. Benefits of public involvement include:

- The resolution of conflicts among different groups during project planning.
- The incorporation of a more comprehensive data base owing to public input.
- More thorough identification and analysis of issues.
- More comprehensive computation of costs and benefits to societal groups.

There has been an evolution of programs for public involvement over the years. In the early days of NEPA activity, the EIS process was dominated by technical issues. Technological solutions were sought for urgent pollution problems, but difficulties stemmed from the general lack of expertise in predicting the effects of these solutions. As better methods were developed for evaluating the direct effects of alternatives, attention was shifted toward secondary impacts. Methods for predicting and evaluating these changes now have become far more sophisticated.

Throughout these early years, public participation programs existed, but they were not always key parts of the planning process. Indeed, many early EISs were prepared after planning decisions had been made. Another factor contributing to weak public involvement was a complacency on the part of the general public. Although many people were concerned about environmental problems and some took very active roles, there still was a widespread willingness to let the experts solve the problems.

For social, economic, and political reasons, this attitude began to change. In the late 1960s, a vocal skepticism about traditional concepts of growth and progress brought about a national commitment to environmental protection. In the early 1980s, growing public concern developed about the need for public participation in project planning. The result is that a larger segment of the population of this country is willing to take, indeed is demanding, a more active role in making decisions about the environment.

Citizens who want and should have a role in the decision-making process often lack familiarity with the technical topics. There is a necessity for programs that will:

- Provide a wide segment of the public with the information they need to participate in planning and decision making.
- Provide this informed public with adequate opportunities and mechanisms for involvement throughout project planning and permitting.

Better methods and materials must continue to be developed so that one can respond to the evolving needs of government and the public. Approaches that will do this are presented in this chapter after a discussion of the regulatory framework that provides for public participation.

4.2   REGULATORY FRAMEWORK

In this section, the regulatory framework that has led to the requirement of public participation programs will be discussed. The general CEQ regulations in that regard will be described and a typical agency approach to their development will then be presented.

4.3   NEPA REQUIREMENTS

Public participation is both implicitly included in the NEPA process and explicitly mandated in the CEQ regulations. Public involvement requirements are specified for all NEPA reviews under 40 CFR Chapter V (Parts 1500–1508), as well as for particular programs. The following actions are required of federal agencies responsible for ensuring NEPA compliance:

- A diligent effort to involve the public in preparing and implementing NEPA procedures.
- Providing public notice of meetings and available documents to:
  1. Specific requestors.
  2. The *Federal Register* for actions of national concern.
  3. State and area wide clearinghouses.
  4. Indian tribes on reservations.
  5. Local newspapers or other local media.
  6. Community organizations.
  7. Newsletters.
  8. Individuals by direct mailing (for local actions).

- Holding public hearings and meetings where there is substantial environmental controversy concerning the proposed action or a request by another agency with jurisdiction over the action.
- Soliciting information from the public.
- Explaining sources of information available for interested persons.
- Making the EIS and supporting information readily available in conveniently located public places, such as libraries.

The following are other references to the public in the NEPA regulations which further support the concept of public involvement in decision making:

- Sections 1500.2(b) and (d) refer, respectively, to making the NEPA process more useful to decision makers and the public, and encouraging and facilitating public involvement in decisions that affect the quality of the human environment.
- Section 1500.4(f) indicates that agencies must emphasize portions of the EIS useful to decision makers and the public.
- Section 1501.2(d) indicates that when an action is planned by a private or nonfederal entity, and the agency can reasonably foresee involvement in the action, then the agency must consult with interested persons even before federal involvement.
- Section 1501.4(e) requires that the agency make a finding of no significant impact (FONSI) available to the public.
- Section 1507.4(a) requires that, as part of the scoping of the issues process, the agency invite the participation of interested persons.
- Section 1503.1(a) requires that the agency request comments from the public, affirmatively soliciting comments from persons or organizations who may be interested or affected.
- Section 1506.2(b) indicates that federal agencies should cooperate with state and local agencies to the fullest extent possible, including joint studies and joint public hearings when this will reduce duplications.

4.4   EPA POLICY

Every federal agency has adopted procedures to comply with the CEQ regulations for public participation in the EIS process. In this section, the EPA procedures will be utilized because they are typical of those required, they are quite comprehensive, and the EPA probably does more EISs, either as lead or cooperating agency, than any other federal agency.

The EPA procedures for implementing NEPA regulations are contained in 40 CFR, Part 6, Subpart D, Public and Other Federal Involvement, which describes in detail how the EPA is to proceed with respect to public involvement. These regulations identify the key components of public participation as: publication of *notices of intent;* conduct of *public meetings* or *hearings;* public review of *findings of no significant impact;* and dissemination of the *record of decision.* Requirements also address how copies of these documents are to be made available to the public.

In 40 CFR, Part 25, the EPA has issued detailed requirements of participation mechanisms for various programs. These regulations present requirements for how public hearings, public meetings, and advisory groups are to be utilized as mechanisms of a public participation program. The regulations describe facilitation of participation in terms of scheduling and conducting hearings. The interdependent nature of all of these mechanisms also is stressed. Public involvement is discussed in terms of achieving balanced participation of all interest groups, which is a key consideration in any public participation process. Requirements call for responsiveness summaries to demonstrate the efforts made at key decision points to facilitate public participation. The agency also must prepare evaluations of the effectiveness of public participation.

The execution of these actions also is required by National Pollutant Discharge Elimination System (NPDES) compliance and permit enforcement regulations. Financial assistance, such as grants, can be made only if public participation regulations are satisfied. For approved state programs, the EPA is required to monitor state compliance with public participation requirements and is empowered to withdraw approval for noncompliant programs.

In 1981, the EPA published a final *EPA Policy on Public Participation* (46 FR 5740), which addresses both the EPA and other government entities carrying out EPA programs. This policy makes an explicit assumption that "agency employees will strive to do more than the minimum required" for public participation. The stated objectives in this policy are to foster increased dialogue between agency officials and the affected parties, early anticipation of conflicts and open discussion of differing opinions among affected parties, and encouragement of mutual trust between government officials and the concerned public. Specific procedures which should be used to achieve these public participation objectives are noted briefly:

- *Identification of parties who may be interested in, or affected by, a proposed EPA project or program*—When the project requires an EIS, the scoping process could be used to identify these parties. Agency officials should develop a contact list and use this list to send notices of hearings, meetings, field trips, or the release of project reports to interested parties.
- *Outreach*—Agency officials must provide policy, program, and technical information to interested parties as early as possible in the planning process. This information must be made available at places that are easily accessible to these interested parties. Efforts should be made to ensure that the public understands the technical aspects of the program. This understanding could be achieved through the publication of fact sheets or technical summaries, surveys, or interviews of community members, public service announcements, news releases, and other educational activities such as workshops and field trips. When announcing public meetings or hearings, agency officials should give a minimum of 45 days notice.

- *Dialogue*—Consultation with interested parties must be undertaken before agency decisions are made. Techniques for increasing dialogue between agency officials and the public include, but are not limited to, citizen advisory committees, workshops, conferences, small group meetings, and toll-free information lines. Any advisory committee formed must present a balance of interests in its membership.
- *Feedback*—Government agencies must provide responses to public inquiries and comments to interested parties regarding the outcome of the public involvement procedures. This feedback must specify the effect that any public comments had on proposed government actions.

The EPA also encourages the *development of public participation work plans* which specify key decisions that are subject to public involvement, staff, and budget resources for participation activities, potential affected parties, and a schedule for public participation activities. The work plans also should identify procedures for conducting the four functions outlined previously: identification, outreach, dialogue, and feedback.

The EPA's public participation regulations and policies call for comprehensive and meaningful public involvement. Since the early 1980s, the agency has been striving to develop even more innovative approaches to public participation. This has been largely due to the fact that public awareness of environmental problems has shifted to hazardous waste sites with increased agency focus on hazardous waste management.

Because of the perception of health risks, agency officials have been striving to develop more open public participation that will earn them the trust of concerned communities. Thus, for example, at each Superfund site where a remedial action is undertaken, a Community Relations Plan is completed which presents a brief case history of the contamination problem and recommendations for public participation activities that the EPA should undertake to address residents' concerns. The underlying rationale for this approach is that if an agency first analyzes the nature and intensity of community concerns for a site-specific project, it will be able to develop a more effective public participation package. More recently, the EPA has advocated developing public involvement plans to specify public participation activities at Resource Conservation and Recovery Act (RCRA) sites. The Act and other sites are discussed in detail later in Chapter 11.

The EPA also has recognized that, even when a comprehensive, sensitive public participation program has been implemented at a hazardous waste site, disputes may still arise between concerned community residents and agency officials. For this reason, the agency has been exploring alternative means for resolving these disputes. For instance, the EPA's Superfund Community Relations Office sponsored a pilot project where professional environmental mediators (i.e., facilitators) were sent to three Superfund sites to address and resolve conflicts that had emerged between the EPA and communities. Based on these three individual cases, the Superfund Community Relations Office developed some general guidelines on how and when the EPA might use this conflict management technique to resolve disputes that have emerged concerning agency actions.

Other divisions within the EPA have used regulatory negotiation in an attempt to incorporate the concerns of competing interest groups into environmental regulations.

The EPA requirements may be summed up in the following specific public participation mechanisms:

Public Notices. A list should be developed of those persons and organizations interested in or possibly affected by proposed activities. Those on the list must receive timely and periodic notification of the availability of materials and early advance notification of public hearings. Public notification must be given at least 30 days prior to major decisions not covered by public meetings or hearings in order to allow time for public response.

Public Consultation. Public consultation and the exchange of views between governmental agencies and interested and/or affected persons and organizations may take several forms. These include public hearings, public meetings, and advisory groups as well as less formal consultation mechanisms (i.e., task forces, workshops, and informal personal communication with individuals or groups). The regulations specifically state that "merely conferring with the public after an agency decision does not meet this requirement." Therefore, information must be distributed in a timely manner during the decision-making process. Public agencies should encourage full presentation of the issues at an early stage so that disagreements can be resolved and responsive decisions can be made.

Public Hearings. Notice of each public hearing must be well publicized and mailed to interested and/or affected parties of record at least 45 days prior to the date of the hearing. If there are no substantial documents to be reviewed or no complex or controversial issues to be addressed, the notice requirement may be reduced to 30 days. The notice must include matters to be discussed and may be accompanied by a discussion of the agency's tentative determinations of major issues (if any), information on the availability of a bibliography of relevant materials (if deemed appropriate), and procedures for obtaining further information. Relevant data, reports, and so on must be available at least 30 days before the hearing. Hearing locations and times must facilitate attendance and a complete record of the hearings must be available for public review. Public meetings are not required to comply with formal hearing regulations, although at least a 30 day notice is required before a public meeting.

Advisory Groups. Advisory groups are required for state, interstate, or local agencies involved in activities supported by EPA financial assistance. Primary responsibility for decision making rests with elected and appointed officials, but all segments of the public must have the opportunity to participate in environmental quality planning. Advisory groups are formed to foster constructive interchange and to enhance the prospect of community acceptance of agency action. Membership of the group should represent a balance of interested parties.

Responsiveness Summaries. When the EPA conducts public participation activities in accordance with 40 CFR Part 25 for grant programs, it must prepare a responsiveness summary at decision points in the grants process. Each responsiveness summary must identify public participation activities conducted, describe the matters on which the public was consulted, summarize the public's views and comments, and outline agency responses to the public.

Special efforts must be undertaken to coordinate public participation activities with those of closely related programs wherever the effort can be enhanced. Hearings and meetings on the same matter may be held jointly. If state permit programs are approved in lieu of federal programs, they must be monitored by the EPA during the annual review of the state's program. The EPA may withdraw an approved program for failure to comply with applicable public participation requirements. As an example, a federal court in Chicago ordered the EPA to withdraw approval of the Illinois NPDES permit program because of EPA's failure to establish guidelines for citizen participation in the enforcement of NPDES programs as required by Section 101(e) of the Clean Water Act (described in detail in Chapter 7). The court ruled that the Part 25 regulations were not sufficient in regard to enforcement.

4.5   APPROACH TO A PUBLIC PARTICIPATION PROGRAM

In the development of a public participation program for an EIS, the following questions should be answered:

- Who is the public affected by the proposed action?
- Where do we find the public?
- What do we want from the public?
- How can mechanisms be provided for input?
- When is input in the study needed?
- How can quality control be assured?

Answers that have evolved to these questions follow:

4.5.1   WHO IS THE PUBLIC AFFECTED BY THE PROPOSED ACTION?

The public is not a unitary mass. Audience segments of the public can be identified in terms of demographic and geographic characteristics such as interest groups, employment categories, income levels, social groups, or locations. A given number of the public often will be included in more than one impacted audience segment. Each segment will have a somewhat different value system, and the function of the public participation program is to uncover the conflicts in values at an early date and provide a forum for their resolution whenever possible.

4.5.2   WHERE DO WE FIND THE PUBLIC?

There are several different methods to locate members of the affected public. A mix of several of these approaches for public participation tends to be optimal.

4.5.2.1   Self-Identification

A citizen or group may inject themselves into the planning process via petition, appeal, public hearing, election, suit, protest demonstration, or publicity. More informal self-identification may be made by correspondence or telephone calls.

4.5.2.2   Group Identification

One may make contact with the public defined by geographic location, interests, or social class. Interest groups can be located by consulting lists of state and national associations. Often these lists are maintained by city and state agencies, university departments, professional associations, or good government groups such as the League of Women Voters. In addition, commercial firms that sell specialized mailing lists can be contacted. Clipping files at local newspapers and libraries are another source of names. Social groups may be located through the public or private agencies that serve them.

4.5.2.3   Third Party Identification

Third party identification is much like group identification, except that it is done by a third party. Possible third parties are:

- A volunteer citizens committee.
- A professional consultant.
- The national association of an interest group.

The same techniques for locating specific members of the public are used by both group and third party identification efforts.

The techniques described thus far make use of historical data. They attempt to find persons already on record as having an interest in the process. In addition, an effort should be made to locate new names. To encourage self-identification, the following techniques may be used:

- Newspaper advertisements.
- Radio and TV spots; public service announcements.
- Establishment of a toll-free hot line.
- Distribution of brochures and other public information materials at sites where interest groups or social groups are likely to congregate.

To encourage third party identification, a snowball interview technique can be used. The public participation specialist begins by interviewing a group of persons known to have an interest and asks them to identify others whom they expect would have an interest. These persons are subsequently interviewed, and the process continues until no new names are forthcoming.

It may be desirable to subscribe to and clip appropriate local newspapers as another source of information from the public. This provides a valuable insight to local issues and allows monitoring of the success of publicity and public involvement measures. New names for mailing lists also can be located from these sources.

4.5.3   WHAT DO WE WANT FROM THE PUBLIC?

There are two objectives when soliciting public input. The first is a short-term objective and consists of information which often includes local perception of issues, description of value systems, confirmation of background facts and figures, review of study findings, and reaction to alternative courses of action. The second objective is the building of positive attitudes toward the agency for whom the EIS is being done and its missions.

4.5.4   HOW CAN MECHANISMS BE PROVIDED FOR INPUT?

The public affected by the outcome of a project is not necessarily well informed about the mission of the federal agency, the purpose of the study, and how it fits into overall regional development. In addition, members of the public may have had little experience with public participation exercises and may need help in overcoming language, cultural, or economic barriers. Mechanisms appropriate for input should accomplish two ends: public education and information gathering.

Public education mechanisms may include:

- Dissemination of pamphlets, newsletters, and newspaper special supplements.
- Placing a display booth at a high-traffic public location.
- Press release or feature story in local media.
- Participation in TV or radio forums such as those presented by educational stations or general purpose talk shows.
- Central depository of interim study findings at local libraries.

Information gathering mechanisms may include:

- Public meeting/public hearing.
- Workshop.
- Telephone hotline.
- Opinion surveys.
- Speakers bureau.

Each mechanism has significant advantages and disadvantages that must be evaluated in relation to time, funding, personalities involved, number of participants, and the level of communications skills of these undertaking the information gathering.

4.5.5 WHERE IN THE STUDY IS INPUT NEEDED?

The earlier in a study that public input is solicited, the greater the likelihood that the study will be completed on schedule and within budget, and will be socially and politically acceptable to the local populace. Therefore, an adequate budget should be allocated for the scoping process. Effort invested in initial problem definition with local officials and other affected parties usually produces substantial benefits such as a clearer understanding of the project needs and an avoidance of unnecessary costs or misguided efforts.

Public input should be encouraged in both a formal and informal manner. Specific forums must be provided for input. The EIS preparer should be in constant touch with those segments of the public who are most interested in active participation during the development and review phases. Specific forums may take the form of citizen review committee meetings, public meetings, or workshops. Project milestones which lend themselves to such forums include the completion of the plan of study and project background task report, alternatives development task report, and alternatives evaluation task report. Public hearings usually are held 45 days after publication of the DEIS.

In addition to the above approaches to public participation, there are two situations that call for public participation in each EIS: one mandatory and the other optional. The mandatory program involves the scoping meeting, usually held about 30 days after the start of an EIS. This meeting includes all of the public agencies that may be interested in the EIS as well as the general public. The purpose of the meeting is to review the work plan for the EIS and to make additions, changes, or deletions. The public agencies stress those areas of concern to them. The general public often has excellent suggestions for items that are highly specific to the EIS and might have been overlooked otherwise.

The second meeting, the optional one, is a public hearing that takes place after the notice of the availability of the draft EIS is published in the *Federal Register*. The meeting is open to both public agencies and to the general public. The purpose of the meeting is to obtain comments on the draft EIS. This is in addition to written comments that any member of the public may make to the draft EIS. The optional factor refers to the decision by the agency sponsoring the EIS as to whether the proposed project is controversial enough to warrant holding a public hearing. In about 90 percent of the cases, the agency decides to hold the hearing.

To summarize, public participation is an essential part of the EIS process. It has the potential to lead to a better project, as well as to improve the possibility of a welcome from the residents of the project area. The public participation program should start as soon as the EIS, itself, does.

5 The Natural Environment: Earth Resources

The earth resources of a site are critical to the stability of proposed construction. Factors to be considered in this portion of an EIS tend to vary in nature and importance as a function of the specific site under consideration. Generally, the following elements are included:

Soils: characteristics, bearing strength, and susceptibility to erosion.
Geology: bedrock, surficial, and seismicity.
Physiography and geomorphology: topography.

Detailed discussions of each of the preceding categories follow. The discussions include methodologies for determining the existing environment, project impacts and, where possible, mitigating measures.

5.1 SOILS

The State of Maryland Department of Natural Resources (DNR) guidelines for environmental impact studies (Maryland DNR, 1974) list the following items to be considered under soils in environmental impact statements:

"1. Describe the soil associations, series, types, and so on that would be affected by the project (names, spatial extents).

2. Discuss physical–chemical parameters relative to each soil series if, and where appropriate, for example,
    a. porosity and permeability.
    b. bulk density.
    c. soil aeration.
    d. water fraction.
    e. mineral composition.
    f. organic content.
    g. pH.
    h. cation exchange.
    i. nitrogen fixation capacity.
    j. microflora and fauna.
    k. erodibility.
    l. depths, profiles.

3. Discuss use classifications relative to each soil series:
    a. engineering (construction) capabilities.
    b. agricultural suitability.
    c. suitability for septic tank type disposal of sanitary wastes."

This very comprehensive listing should be considered in describing the affected soils in an EIS. Some of the items may be unknown and/or unnecessary, for example, items (h), (i), (j), but they should be included if the information is available.

The primary step in soils investigations is the determination of their general suitability, based on their physical and chemical properties, for projected activities such as structural foundations, roads, pipelines, or land disposal of wastewater or sludge. In practice, the investigations of soils for engineering and environmental properties are carried out in a coordinated program that includes description of the topography and geology of the study area. Faulty conclusions are possible if soils are assessed as suitable for waste application or other pollutant loading processes without considering underlying rock porosity, type, orientation and structure, and degree of fracturing.

A typical approach to soils inquiries includes the following steps:

* Acquisition and review of pertinent published literature and open file reports from the U.S. Geological Survey (USGS), state geological surveys, the U.S. Natural Resources Conservation Service (NRCS), universities, and other local sources.
* Acquisition and evaluation of recent aerial photos.
* Preliminary evaluation of the soils to identify potential impacts and to determine whether reconnaissance fieldwork is necessary.
* Field investigation, if necessary, to verify published information and to rectify data deficiencies.
* Summarization of information on existing conditions, incorporating all data obtained from literature reviews and field surveys.
* Identification and evaluation of potential impacts or hazards (steep slopes, poor soil characteristics, or faults), as well as the existence of prime and unique farmlands and/or minerals.

Existing background data must be collected and reviewed to determine the effects that new or expanded facilities will have on the rate of soil erosion. The NRCS is the primary source of detailed soils information. It sponsors numerous cooperative programs with state universities, counties, and local soil and water conservation districts. Published and unpublished soils maps are available for most areas. Additional or original data may be obtained from metropolitan, regional, or state agencies and from other federal and state agencies for forests, recreation areas, and refuges. If data are not available for a specific case, the NRCS can be commissioned to perform work through a cooperative agreement with the EPA, or qualified EIS personnel can undertake original soils mapping.

Erodibility is an important soils characteristic relative to new developments. During the evaluation of construction-related effects on soils, the potential for erosion must be examined. Soil erosion problems affect clearing and excavation at the facility site, along pipeline and transmission corridors, and other ancillary facility areas, and may necessitate the use of mitigative measures to minimize adverse effects. Mitigations range from the simple placement of hay dikes for containment of water and soil to the construction of elaborate impoundments and drainage ditches. These include seeding, netting, mulching, scarification of exposed soils, and covering of excavated piles of soil. These measures reduce the environmental damage that results from the loss of natural upper soil horizons and the surface water pollution that occurs as a result of sedimentation.

The analysis of erosion potential and slope stability at a specific site generally entails the preparation of a map with overlays that show:

- Unstable terrain features identified on recent aerial photographs and topographic maps (generally at the 1 : 24,000 scale).
- Unstable soil types identified by the NRCS and delineated from soil survey maps, if available.
- Unstable slopes, landslides, escarpments, highwalls, or other unstable features delineated from state and USGS data, if available.
- Unstable slopes or other suspect features delineated from maps or other data supplied by the permit applicant or grantee.

The assembled data are evaluated for their relevance, completeness, reliability, and ramifications. On the basis of this initial assessment, specific areas are targeted for on-site investigations, if necessary, to characterize adequately the stability and erosion potential of selected areas.

In the examination of earth resources, more effort is usually expended upon soils than upon any other single area. This is because

1. The nature of the soils will directly affect the feasibility of the construction of the project.
2. The soil structure is the key factor in the determination of the possible contamination of groundwater.

Soil is a result of the interaction of soil-forming processes on materials deposited by geologic agents. The properties of the soil at any given place are determined by five factors:

1. The physical and mineralogical composition of the parent material.
2. The climate under which the soil material has accumulated and has existed since accumulation.
3. Living organisms on and in the soil.
4. The topography.
5. The length of time the forces of soil formation have acted on the soil materials.

Exhibit 1 shows the results of a typical soil study in an environmental impact study (BREGMAN & COMPANY, 1990).

EXHIBIT 1
Summary of Predominant Soil Characteristics

Site Characteristics	Site A	Site B	Site C
Principal soils	Morley	Elliott, Ashkum, Varna	Bryce, Milford
Drainage	well drained	poorly drained	poorly drained
Slopes (%)	2–4	0–4	0–2
Percolation rate (in./h)	0.63–2.0	0.6–2.0	0.2–2.0
Shrink-swell potential	low	low to moderate	moderate
Depth to early spring water table (in.)	36	0–36	0–24
Potential shallow excavation problems	wetness	wetness	wetness
Hydric soils present	no	yes	yes
Stable bank cuts	yes	yes	yes

Conclusions reached in that study were as follows:

"A summary of predominant soil characteristics for the three sites shows that Site A soils are less wet in late winter and early spring than are the soils at the other two sites. Sites B and C have some hydric soils, the land is flatter, drainage is decreased, and the apparent water table in late winter and early spring is nearer the ground surface. Bank stability in shallow excavations is reported to be satisfactory at all sites and conversations with those in the grave digging and well drilling professions report no water problems in such excavations at all sites."

Examination of the suitability of soils for use as farmland was mentioned earlier in this section. This is a particularly critical item.

The Farmland Protection Policy Act, Pub. L. 97-98 (as of January 16, 1996), has as its purpose to minimize the extent to which federal programs contribute to the unnecessary and irreversible conversion of farmland to nonagricultural uses. It also assures that federal programs are administered in a manner that, to the extent practicable, will be compatible with state, local government, and private programs and policies to protect farmland. This Act states that, "(t)his subtitle does not authorize the Federal Government in any way to regulate the use of private or non-Federal land, or in any way affect the property rights of owners of such land." The Act pertains to prime, unique, and statewide or locally important farmland. Unique is defined as ". . . used for production of specific high-value food and fiber crops. . . ."

Pursuant to this Act, the NRCS promulgated 7 CFR 658. The preamble to this rule states, "(n)either the Act nor this rule requires a Federal agency to modify any project solely to avoid or minimize the effects of conversion of farmland to nonagricultural uses. The Act merely requires that before taking or approving any action that would result in conversion of farmland as defined in the Act, the agency examines the effects of the action using the criteria set forth in the rule, and if there are adverse effects, consider alternatives to lessen them. The agency would still have discretion to proceed with a project that would convert farmland to nonagricultural uses once the examination required by the Act has been completed."

The federal rule provides criteria for establishing a numeric score to the value of a farmland parcel. The highest attainable score is 260 points, and 7 CFR 658.4 provides guidelines for use of the criteria. Once this score is computed, the U.S. Department of Agriculture (USDA) recommends:

> "(1) Sites with the highest combined scores be regarded as most suitable for protection under these criteria and sites with the lowest scores, as least suitable.
>
> (2) Sites receiving a total score of less than 160 be given a minimal level of consideration for protection and no additional sites be evaluated.
>
> (3) Sites receiving scores totaling 160 or more be given increasingly higher levels of consideration for protection.
>
> (4) When making decisions on proposed actions for sites receiving scores totaling 160 or more, agency personnel consider:
>
>   (i) Use of land that is not farmland or use of existing structures.
>
>   (ii) Alternative sites, locations and designs that would serve the proposed purpose but convert either fewer acres of farmland or other farmland that has a lower relative value.
>
>   (iii) Special siting requirements of the proposed project and the extent to which an alternative site fails to satisfy the special siting requirements as well as the originally selected site."

A number of Midwestern states use comparable systems. For example, the state of Illinois has an Land Evaluation and Site Assessment (LESA) System (Illinois Department of Agriculture, 1992). Under this system, the NRCS performs the land evaluation portion of the assessment. This is based principally upon the productivity of the soils for agricultural use on a scale of 0 to 100 with 100 representing the best agricultural land. This portion of the assessment is the same as the rating system used by the USDA. The Illinois Department of Agriculture then performs the site assessment of the land parcel based upon 16 criteria for a total of 200 maximum points. This assessment considers all factors relative to a specific parcel of land, other than soils, which would further determine the viability of a site for agricultural or nonagricultural use. Thus, under LESA there are 300 total rating points with which to judge the relative value of the farmland, whereas under the USDA system there would be 260 total rating points for the same purpose. The site assessment criteria take into consideration the proximity of the parcel to a city, availability of a central water system, availability of a central waste disposal system, size of site, adjacent land use, the percentage of the site in agriculture, the percentage of adjacent land in agriculture, and similar site evaluation factors.

These numbers differ when a corridor is under consideration. In that case, the land evaluation is worth 150 points and the site assessment also is worth 150 points. The Illinois LESA System applies a 225 cutoff point when evaluating state and federally funded projects. Site or corridor alternatives receiving 175 or fewer points have a low rating for protection, and it is not necessary to evaluate additional alternatives. Those alternatives receiving 176 to 225 points are in the moderate range for protection, and at least one build alternative should be considered. In most cases, alternatives exceeding the 225 point level should be retained for agricultural use, and an alternate site should be utilized for the intended project. Selecting the alternative with the lowest total points will usually protect the best farmland located in the most agriculturally viable areas.

5.2   GEOLOGY, PHYSIOGRAPHY, AND GEOMORPHOLOGY

Geologic conditions are significant in EIS preparation because they can place constraints on the nature, design, or location of the proposed activity, as well as determine the impacts which the activity will have on other resources. The EIS must consider the nature and configuration of both the surface and subsurface materials that occupy the project area.

The state of Maryland guidelines recommend inclusion of the following information (Maryland Department of Natural Resources, 1974):

"Geology:

1. Bedrock—identify to extent possible:
   a. Formations (names, depths, thicknesses, rock types, extents, mineral composition, etc.).
   b. Economically valuable minerals (sand and gravel, clay, etc.).
   c. Fossil deposits.
   d. Unique or limited minerals, or formations, etc.
   e. Foundation conditions.
   f. Fault zones, karst topography (sinkholes, springs, subsurface drainage, etc.).
2. Surficial:
   a. Consider (a)–(f) above.
3. Marine–Estuarine:
   a. Consider (a)–(e) above.
   b. Consider sediment quality parameters of sediment types.
      1) Percent organic content, percent sand, bulk density, and so on.
      2) Heavy metals (Cr, Pb, Hg, etc.).
      3) Metals present (Fe, Cu, Zn, Mn, etc.).
      4) pH.
      5) Salinity.
      6) Cation exchange capacities.
      7) Exotic contributions (oils, pesticides).
      8) Sulfides.
      9) Nitrogen.
      10) Inorganic carbon.
      11) Organic compound.
      12) Putrescibility index (odor).
   c. Sediment quantity—rates of erosion, methods, and sites for disposal, and so on.

Physiography and geomorphology:

1. Discuss impact upon geomorphologic features (scarps, terraces, ravines, depressions, beaches and sand source areas, dunes, streams, drainage patterns, berms, sills, dikes, islands, shoal areas, erosion areas, and rates of erosion, etc.).
2. Describe changes in elevations and slopes.
3. Discuss impact upon man-made structures (gravel pits, road cuts, borrow areas, mounds, drainage ditches, fill areas, etc.)."

The topography of an area is considered as part of an overall approach to physiographic information. This information generally includes the underlying geologic structure, drainage patterns, total and local topography relief, and the nature of the soils. Topographic information or stereo aerial photography is used to evaluate facility sites and gradients, to determine optimal orientation, and to identify areas that may require soil stabilization based on number and placement of road and stream crossings, volumes of fill and excavations, and effects of these actions on natural drainage.

Geological data for the first stage of evaluation are available in the form of local or regional geologic maps from the USGS and/or state geological surveys, NRCS, universities, and other local sources. These data indicate the general nature of the bedrock, probable depth to significant rock strata, and the presence of any known mineral deposit.

Topography, geomorphology, and geology are usually described in the NEPA document in terms of unique features, slope stability, shallow bedrock areas, aquifer recharge areas, subsidence, seismic considerations, and special features. An investigation will be made to determine what, if any, geological and topographical features could constrain development at each alternative site or contaminate groundwater and what additional information may be necessary. Recent aerial photographs are useful. Fieldwork, if necessary, is undertaken to verify published information and rectify data deficiencies.

Evaluation of the impacts of the project on the geological environment generally is needed only when major rock excavation and/or mining are contemplated. Geological field reconnaissances are undertaken primarily to verify data from the literature, to orient staff investigators to the area, and to identify deficiencies in the data base. If required, geological field work would involve construction of detailed geological maps and cross-sections, measurement of the thickness and orientation of bedrock strata, and petrographic analysis of rock samples.

Major geological constraints to new construction that may be evident at the initial level of analysis are as follows:

• The bedrock is too hard for excavation or too weak to support proposed structures.
• The mineral deposits would be destroyed or occluded by development of the project.

- The existence of hazardous structures (active faults) may jeopardize the project.
- Shallow depth to bedrock may preclude the use of soil absorption systems for disposal of treated wastewater.

Seismicity usually is examined as a part of the geological study in a NEPA report. Seismic risk related to structural damage in the United States may be represented by zones having a relative scale zero through four, with Zone 0 not expected to encounter earthquake damage and Zone 4 expected to encounter the greatest risk (see Exhibit 2). Building codes require that the design and construction of a building comply with the requirements for the seismic zone in which the building is located, with the required strength increasing as the zone classification goes from zero to four. For example, a building in San Francisco would need to have several times the earthquake resistance that a similar building in Wisconsin would require.

5.3   MITIGATING MEASURES SUMMARY

In general, mitigating measures in the case of earth resources are almost entirely preventive in nature, similar to the case with groundwater. The earth resources analysis serves primarily to direct the location of the project to one that is suitable from a geologic viewpoint and away from a dangerous location. The detailed soils study goes together with the groundwater examination in determining how best to minimize or eliminate groundwater contamination.

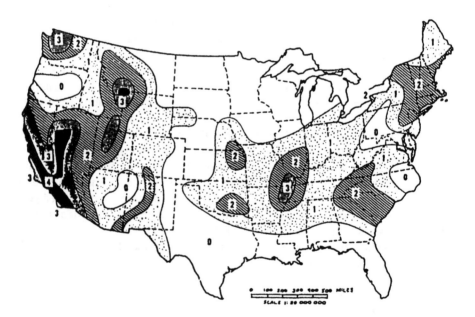

Exhibit 2   Seismic zones in the contiguous 48 states. (From the BOCA National Building Code, 1987. With permission.)

REFERENCES

BOCA National Building Code, 1987.

Environmental impact study of proposed national cemetery sites in northeast Illinois, BREGMAN & COMPANY, Inc., 1990.

Land evaluation and site assessment system, Illinois Department of Agriculture, 1992.

Revised guidelines for implementation of the Maryland Environmental Policy Act, Maryland Department of Natural Resources, June 15, 1974.

6 The Natural Environment: Biology

In considering the effects of a proposed project or program on the biology of an area, one must examine all of the possible entities in all locations—air, land, and water. These locations, in turn, may be subdivided into factors such as deserts, tundras, wetlands, coastal areas, fresh and salt water, and so on.

Both animal and vegetable matter have to be taken into account. The various families and taxa in each of the categories that may be present in the area under study must be examined. A broad and incomplete listing that illustrates this point follows:

Water	Air	Land
algae	birds, bats, etc.	vegetation
phytoplankton	flying insects	plants
zooplankton		shrubs
benthic organisms		trees
fish at all stages		snakes
		insects
		small mammals
		large mammals

A comprehensive listing of all of the biological entities that may require consideration in an EIS has been prepared by the Maryland Department of Natural Resources (Maryland DNR, 1974) as follows:

"1. Vegetation and Flora*
    a. vascular marine and non-marine vegetation and flora
        1) vegetation types—wetland, bottomland hardwood forest, pine–oak forest, pine forest, sweetgum–beech forest, fields, etc.
            a) spacial extent (delineate acreage)
            b) floral composition (species list)
            c) vegetative structure (density, cover values stratification, successional stages, etc.)
            d) diversity (equitability component [evenness], richness component [number of species], and plant species diversity index or foliage height diversity index)
            e) edge/area ratio
            f) economically important species
        2) system maintenance functions
            a) productivity values of types
            b) food chain relationships (energy transfer)

---

*The term vegetation deals with plant communities while the term flora is concerned with individual taxa (genera, species, etc.), that is, a list of species present.

  c) detritus production
  d) reproductive success and dispersal rates
  e) value for animal habitat
  f) oxygen production
  g) nutrient cycling capacity
  h) flood buffer capacity
  i) erosion control capacity
  j) sediment entrapment function
  k) turbidity reduction capacity
  l) water pollution abatement capacity
  m) firebreak function
  n) salt water buffering capacity
  o) water storage capacity
3) rare or endangered species, populations and communities
4) unique or virgin stands and habitats, large den trees, large or record tree specimens, etc.
5) pestiferous or parasitic species
6) delineate and describe agricultural/horticultural/forestry crops
b. non-vascular, non-marine vegetation and flora (mosses, lichens, fresh water phytoplankton, etc.)
  1) vegetation types—upland, lowland, marsh, etc.
    a) spacial extent
    b) temporal distribution
    c) floral composition (species list)
    d) vegetative structure (density, stratification, etc.)
    e) diversity (equitability component [evenness], richness component [number of species] and plant species diversity index)
    f) economically important species
  2) system maintenance functions
  3) rare or endangered species, parasitic species, pestiferous species, unusual vegetation types, etc.
c. non-vascular, marine vegetation and flora (phytoplankton, etc.)
  1) vegetation types—estuarine phytoplankton, benthic algae, etc.
    a) spacial extent
    b) temporal distribution
    c) floral composition (species list)
    d) vegetative structure (density, stratification, etc.)
    e) diversity (equitability component [evenness], richness component [number of species] and diversity index)
    f) economically important species—especially for finfish and shellfish industry
  2) system maintenance functions
    a) phytoplankton productivity
    b) turnover rates
    c) oxygen production capacity
    d) food chain relationships (energy transfer)
2. Faunal Communities and Fauna*
  a. Discuss impact upon spacial and temporal distribution of faunal land communities (wetland, field, forest, dune, beach, etc.) including:

1) construct a species list
2) delineate habitats and niche requirements
3) discuss predator–prey relationships
4) discuss reproductive success, dispersal rates, migrations (seasonal, immigration, and emigration rates)
5) consider population natality, mortality, longevity, etc.
6) cicadian rhythms
7) nocturnal activities
8) population growth rates and size fluctuations (seasonal, annual, etc.)
9) turnover rates, densities and trophic levels
10) age distributions
11) feeding, shelter, nesting and wintering, etc. areas and migration routes
12) behavior activities (territoriality, feeding behavior, physiological stress, community homeostasis, mating rituals, symbiotic relationships, etc.)

b. Discuss impact upon spatial and temporal distribution of aquatic faunal communities (marine, estuarine, and fresh water) such as intertidal flats, shellfish reefs, limnetic and littoral communities, benthic communities, etc.)

c. Discuss any past atypical animal mortalities (fish kills, waterfowl cholera, botulism, MSX virus, etc.)

d. Discuss any economically important or valuable species or populations potentially affected by the project such as:
1) commercial species (finfish, shellfish, bloodworms, crabs, etc.)
2) game or sport species (shellfish, finfish, waterfowl, upland game, big game, etc.)

e. Discuss pestiferous species such as:
1) mosquitoes, tabanid flies, agricultural or horticultural crop pests, forestry pests, shellfish predators, etc.
2) parasitic species (upon plant, wildlife, and humans)

f. Describe any rare and endangered species, populations, communities and habitats that could potentially be disturbed or displaced by the project."

6.1 LEGISLATION

There are a number of pieces of legislation, possible violations of which must be considered in examining impacts of a proposed project or program on the biological setting for an area. Some of these are discussed elsewhere in this book, including the very important Clean Water Act. Some of the others that are the most pertinent are described briefly below.

6.1.1 THE ENDANGERED SPECIES ACT

Under the Endangered Species Act, federal agencies are prohibited from jeopardizing the continued existence of threatened or endangered species or destroying or

---

*The term faunal communities refers to populations or aggregations of populations while the term fauna is concerned with individual taxa (genera, species, etc.), that is, a list of species present.

adversely modifying habitats that are essential for the survival of these organisms. Survey and mitigation measures are required.

In each EA and EIS, a literature review and field survey should be conducted of the rare, threatened, endangered species which occur or which could potentially occur on the project site. The literature review is conducted for the potential occurrence of both state and federally listed species of plants and animals. This is augmented by maintaining contacts with the appropriate state and federal agencies. An on-site survey then is made, if required, in order to confirm the potential occurrence of these species on the site. If any protected species are found on the site, their location and habitat are noted. An estimate of the potential occurrence of both plant and animals on the site also is made based on the results of the field vegetation survey and literature review. This analysis is required because of the highly mobile nature of many animals and because plant species are typically highly seasonal in their occurrence.

Additional literature review is conducted if a species is found to actually occur on the site or if the site has a very high potential as a habitat. The nature of these species is described, as well as the potential impacts of the project. If the site harbors protected species, emphasis is placed on designing the facilities and structures which are part of the project in such a manner as to minimize potential impacts. Early coordination is undertaken with the EPA and relevant state and federal agencies, as needed, in order to achieve the best design which will minimize or completely avoid effects on rare, threatened, or endangered plants and/or animals. If federally protected species are involved, one begins the informal coordination process with the U.S. Fish and Wildlife Service at the appropriate time and after consultation with the EPA. However, by incorporating careful site design and other techniques, involvement of both state and federal agencies, with respect to rare, threatened, and endangered species can be minimized and delays in the project avoided. At the same time, effective communication with the agencies of concern ensures that the health and well-being of any potentially affected species are protected.

6.1.2   FISH AND WILDLIFE COORDINATION ACT

The Fish and Wildlife Coordination Act was first passed in 1934 and has been amended a number of times. The Act, in its declaration of purpose, states that:

> " . . . wildlife conservation shall receive equal consideration and be coordinated with other features of water-resource development programs through the effectual and harmonious planning, development, maintenance, and coordination of wildlife conservation and rehabilitation . . . to provide assistance to, and cooperate with, Federal, State, and public or private agencies and organizations in the development, protection, rearing, and stocking of all species of wildlife, resources thereof, and their habitat, in controlling losses of the same from disease or other causes, in minimizing damages from overabundant species, in providing public shooting and fishing areas, including easements across public lands for access thereto, and in carrying out other measures necessary to effectuate the purposes of said sections."

The Act then goes into detail on how federal agencies are to work together to achieve the preceding objectives. It goes on to say flatly that:

". . . whenever the waters of any stream or other body of water are impounded, diverted, the channel deepened, or the stream or other body of water otherwise controlled or modified for any purpose whatever, including navigation and drainage, by any department or agency of the United States, adequate provision, consistent with the primary purposes of such impoundment, diversion or other control shall be made for the use thereof, together with an area of land, water, or interests therein, acquired or administered by a Federal agency in connection therewith, for the conservation, maintenance, and management of wildlife resources thereof, and its habitats thereon, including the development and improvement of such wildlife resources. . . ."

The Act defines wildlife as ". . . birds, fishes, mammals and all other classes of wild animals and all types of aquatic and land vegetation upon which wildlife is dependent."

When EISs are examined by the Fish and Wildlife Service, particular attention is paid to ensuring that the proposed project will not adversely affect fish and wildlife. If there is a negative effect, suitable mitigating measures must be employed.

6.1.3 COASTAL ZONE MANAGEMENT ACT

The preamble to this 1968 Act states the Congress finds that . . .

"(c) The increasing and competing demands upon the lands and waters of our coastal zone occasioned by population growth and economic development, including requirements for industry, commerce, residential development, recreation, extraction of mineral resources and fossil fuels, transportation and navigation, waste disposal, and harvesting of fish, shellfish, and other living marine resources have resulted in the loss of living marine resources, wildlife, nutrient-rich areas, permanent and adverse changes to ecological systems, decreasing open space for public use, and shoreline erosion.

(d) The coastal zone, and the fish, shellfish, other living marine resources, and wildlife therein are ecologically fragile and consequently extremely vulnerable to destruction by man's alterations.

(e) Important ecological, cultural, historic, and aesthetic values in the coastal zone which are essential to the well-being of all citizens are being irretrievably damaged or lost."

Amendments to the Act in 1980 required the development of special area management plans (SAMP) for areas of the coastal zone considered to be of particular importance. SAMP are comprehensive plans that provide for natural resource protection and reasonable coastal-dependent economic growth containing a detailed and comprehensive statement of policies; standards and criteria to guide public and private uses of lands and waters, and mechanisms for timely implementation of the designated geographic areas. They are also intended to provide for increased specificity in improved protection of life and property in hazardous areas, including those areas likely to be affected by land subsidence, sea level rise, or fluctuating water levels of the Great Lakes, and improved predictability in governmental decision making.

As a result of this Act, programs to protect the coastal zone have been developed by each of the states that have coastal areas. These state programs are to provide for:

1. The protection of natural resources, including wetlands, flood plains, estuaries, beaches, dunes, barrier islands, coral reefs, and fish and wildlife and their habitat, within the coastal zone.
2. The management of coastal development in flood-prone, storm surge, geological hazard, and erosion-prone areas, and in areas likely to be affected by or vulnerable to sea level rise, land subsidence, and saltwater intrusion, and by the destruction of natural protective features such as beaches, dunes, wetlands, and barrier islands.
3. The management of coastal development to improve, safeguard, and restore the quality of coastal waters, and to protect natural resources and existing uses of those waters.
4. Priority consideration to coastal-dependent uses and orderly processes for siting major facilities related to national defense, energy, fisheries development, recreation, ports and transportation, and the location, to the maximum extent practicable of new commercial and industrial developments in or adjacent to areas where such development already exists.
5. Public access to the coasts for recreation purposes.
6. Assistance in the redevelopment of deteriorating urban waterfronts and ports, and sensitive preservation and restoration of historic, cultural, and esthetic coastal features.
7. The coordination and simplification of procedures in order to ensure expedited governmental decision making for the management of coastal resources; continued consultation and coordination with, and the giving of adequate consideration to the views of, affected federal agencies.
8. The giving of timely and effective notification of, and opportunities for public and local government participation in, coastal management decision making.
9. Assistance to support comprehensive planning, conservation, and management for living marine resources, including planning for the siting of pollution control and aquaculture facilities within the coastal zone, and improved coordination between state and federal coastal zone management agencies and State and wildlife agencies.
10. The study and development, where appropriate, of plans for addressing the adverse effects upon the coastal zone of land subsidence and of sea level rise.

Any EIS for a project that may impinge upon a coastal zone must consider ways of complying with state regulations. This is especially true if a state has an approved coastal zone management program through the Office of Coastal Zone Management, National Oceanic Atmospheric Administration (NOAA). Federal agencies with development projects within the coastal zone, including civil work activities, must assure that those activities or projects are consistent to the maximum extent practicable with the approved state program.

6.1.4   MARINE PROTECTION, RESEARCH, AND SANCTUARIES ACT
(OCEAN DUMPING)

The Marine Protection, Research, and Sanctuaries Act of 1972, better known as the Ocean Dumping Act, regulates the ocean dumping of all types of materials that may adversely affect human health, the marine environment, or the economic potential of the oceans. The EPA is authorized to designate sites where ocean dumping may be permitted or prohibited and to issue permits for material other than dredged materials. The U.S. Army Corps of Engineers is responsible for issuing permits to dump dredged material at sites designated by the EPA.

The Act prohibits the dumping of radiological, chemical, and biological warfare agents and high-level radioactive wastes. For other wastes, the EPA must determine through applicable criteria that their permitting for dumping will not unreasonably degrade or endanger human health, welfare or amenities, or the marine environment, ecological systems, or economic potentialities.

In establishing criteria to regulate ocean dumping, the Act requires that consideration be given, but not necessarily limited to:

* The effect of such dumping on fisheries resources, plankton, fish, shellfish, wildlife, shore lines, and beaches .
* The effect of such dumping on marine ecosystems, particularly with respect to the transfer, concentration, and dispersion of such material and its by-products through biological, physical, and chemical processes; potential changes in marine ecosystem diversity, productivity, and stability; and species and community population dynamics.

The EPA issues an EIS on each site designated for ocean disposal. Following the EIS, the site is approved and then a permit is issued for dumping.

6.2   INFORMATION DEVELOPMENT

In determining the preproject state of the natural biological environment, one may utilize existing data, gather new data from the field, or do both. When the time frame is short, as in the case of an EA, then reliance on data in the literature or data previously gathered by other groups or agencies becomes almost imperative. In these cases, there usually is very little in the way of time or funds available for the gathering of field data. On the other hand, for an EIS, and especially for one where the proposed project may impact wetlands or rare or endangered species, the time and funds to gather current field data properly must be taken. In many cases, this may mean data relating to the same species taken at the same locations in each of the four seasons of the year and then a fifth sampling to check the first season for any substantial change that may have occurred in that species during the year of observation.

The impacts on biota are of considerable importance in a NEPA document because they may affect the entire food chain. In case of water, effects start with plankton or benthic organisms, goes through fish, and thence to man. The food chain involving vegetation and animals may follow a similar procedure, so that toxic chemicals that are taken up by small organisms may ultimately become concentrated in man and in some of the higher animals, for example, bald eagles.

In general, the approach to biology in an EIS, starting with an assessment of the existing conditions and then determining what the project impacts will be, includes the following:

- Assess existing literature and information concerning the organisms and environment in the study area and surrounding region.
- Determine the need for field studies.
- Design a qualitative and/or quantitative field sampling program, if needed.
- Evaluate the predicted primary and secondary impacts of the proposed project and alternatives based on the examination of all available data.

For many projects, data are readily available that can be used to describe the existing baseline conditions. Such information is normally acquired from the following sources:

- U.S. Fish and Wildlife Service.
- USDA Natural Resources Conservation Service.
- U.S. Army Corps of Engineers.
- EPA.
- State departments of natural resources.
- Colleges and universities.
- Interest groups (Audubon Society, Nature Conservancy, etc.).
- Contacts with federal, state and local officials.

Where additional information is necessary, aquatic and terrestrial field surveys are conducted as described below.

6.3   TERRESTRIAL BIOTA

The methodologies for assessing impacts to terrestrial biological resources from most types of development projects (power generation facilities utilizing coal or other fossil fuels, pulp and paper mills, hydroelectric facilities, chemical plants, refineries, waste treatment facilities, etc.) are similar. Existing information must be collected and reviewed to establish if sufficient data are available to be presented in a form suitable for the evaluation of impacts. If adequate baseline information is available in-house or from other suitable information sources, existing conditions can be characterized and potential effects from proposed project activities on the biological resources can be discussed in a timely manner.

The analysis presents discussions of major terrestrial communities, including important species and their roles and functions in the system. Detailed species lists of flora and fauna are referenced or appended. The EIS authors also should consult with state and federal fish and wildlife agencies to determine the possible presence and locations of rare and endangered or threatened plant and/or animal species, to estimate project effects on sensitive habitats such as wetlands or other sensitive natural features, and to identify the types and timing of effective measures to minimize adverse impacts. Section 7 coordination with the U.S. Fish and Wildlife Service and Section 404 wetland determinations with the Army Corps of Engineers are often important factors in these evaluations.

If adequate data are not available, the deficiencies are identified and a determination is made as to what additional information is necessary. The following sections outline typical methods for identification of key issues, characterization of existing resource conditions, and evaluation of impacts related to terrestrial biota.

6.3.1 UPLAND PLANT COMMUNITIES

If no recent vegetation map is available, vegetation of the project region is mapped by the interpretation of aerial photographs and/or by field inspections.

A narrative description is prepared for each type of major plant community. The appearance and structures of the community are described and a list of dominant plant species in each layer of the vegetation is presented. The successional status of each community is identified. The analysis focuses on the relative suitability or unsuitability for development of different sections of the project region insofar as vegetation is concerned. Any areas supporting endangered or otherwise unique plants are described and located on a map.

Based on the information described previously, an evaluation can be made of the impacts of the proposed industrial facility.

At the conclusion of the identification phase, the following sensitive types are depicted graphically and discussed:

- Areas where endangered and threatened plants may occur.
- Wetlands (Army Corps of Engineers designated and nondesignated).
- Coastal zone area (where applicable).
- Remnant and relict botanical areas.

6.3.2 BOTTOMLAND/WETLAND PLANT COMMUNITIES

There are four federal agencies with jurisdictional interest in wetlands. They are the Army Corps of Engineers, the EPA, the Fish and Wildlife Service, and the Natural Resources Conservation Service. Definitions of wetlands have been developed by each agency. The Fish and Wildlife Service's definition encompasses both vegetated and nonvegetated areas; definitions for the other agencies include only areas that are vegetated under normal circumstances. All definitions have three basic elements for identifying wetlands: they are hydrology, vegetation, and soil characteristics. Basically, the Fish and Wildlife Service's definition includes mud flats, sand flats, rocky shores, gravel beaches, and sand bars. The Natural Resources Conservation Service emphasizes a predominance of hydric soils in the wetlands definition.

If existing wetlands data are inadequate or conflicting, the group doing the EIS should examine aerial photographs and conduct field checks to further define their extent and nature. Exact sites for proposed construction activities undergo additional field examination to ensure that wetlands are subjected to minimal alteration or none at all.

Plant communities in wetlands can be affected directly and indirectly during the construction phase of a project. Furthermore, severe impacts can result from construction activities on adjacent lands if erosion and subsequent runoff damage the quality of the wetland area irreversibly.

Secondary impacts can result from the proximity of induced development to the wetland area. Impacts caused by urban/industrial nonpoint runoff or additional persons utilizing the area can disturb or essentially destroy vegetation in wetland areas. The loss of wetland areas could also impact the quality and quantity of surface water and groundwater.

Impacts of multiple projects on wetlands are critical. The removal of such land is almost irreversible. Consequently, many coastal states have placed severe limitations on projects that will use up wetlands, whether or not newly created ones are offered in trade.

6.4   WILDLIFE

Amphibians, reptiles, birds, and mammals normally are included in an existing conditions inventory for a NEPA study. In some instances, insects, soil organisms, or other terrestrial life forms also are considered.

An inventory can be performed at different levels of detail, depending on the availability of previously collected and/or published information; the recognition of a significant environmental component, such as the presence of an endangered species and critical habitat for such a species; funding or time restrictions; and other factors. In general, the inventory begins with a review of available literature to obtain descriptions of the project area and of the important species that may inhabit or use the area. Parks or wildlife refuges located in or near the project area also are noted.

The major sources consulted for such information include in-house reference materials; computerized information bases; public, private, college, university, and museum libraries; federal, state, regional, county, and local agency files and publications; state academies of science; park managers and area game biologists; local arboreta, nature centers, zoological societies, state or national Audubon chapters and environmental groups; biology departments at nearby colleges or universities; and local citizens knowledgeable about wildlife populations, such as hunters, trappers, and birders. Many government publications are available for use in depository libraries. Masters' theses usually can be obtained by standard interlibrary loan procedures, and doctoral dissertations can be purchased in hardcover or on microfilm from several sources.

The first step in the impact evaluation process is identification of endangered and threatened species that may be present in the project area. Lists of designated or proposed species are obtained from the U.S. Department of the Interior, Fish and

Wildlife Service, and the appropriate state fish and game agencies. The presence of these species is usually documented for a specific project area in field studies. If one or more federal or state endangered or threatened species potentially is present in the project area, a description of the habits, habitat requirements (food plants or prey, vegetation cover requirements for shelter or breeding, and the size of territory), and tolerance to pollution and human activity is prepared so that potential impacts on these species or their habitats can be determined. If no list of species is available from federal, state, or local sources for the project area or its environs, a working list is compiled by examination of field guides for each group of animals and identification of those whose ranges are included in the project area. Many of these publications also provide information on the habitats and habits of these species. Information on the distributions and abundance of each species is obtained from publications for specific regions, states and localities.

Animal habitats in the project area are identified as part of the vegetation survey, which normally includes the preparation of a land cover map. The species that may be present in these habitats are estimated by the use of literature sources. This correlation is confirmed through field reconnaissance of the project site. The value and utility to wildlife or plant species that comprise these habitats are estimated with the aid of general references.

By means of telephone calls and letters to the knowledgeable agencies and private sources listed previously, additional information is collected in such forms as research reports or game harvest figures from state biologists, and checklists of species from parks, forest preserve districts, or environmental groups. Previous environmental impact statements, research reports, and technical bulletins can be obtained on loan from state departments of natural resources or conservation departments, even if these documents are out of print. Often the sources contacted will know of additional persons conducting research in the area (such as university faculty or students), and of scientists currently preparing publications containing information valuable to the EIS. They also may provide information on the existence of previous environmental reports or studies on nearby areas.

The species lists and text materials prepared for each group of animals includes identification of the species considered to be endangered or threatened at the state and/or federal level, other especially valuable species, and those species known or expected to breed in the project area. The land cover map prepared in the vegetation survey serves as the source of habitat-type information for correlation of each species with the habitats that it uses. Information on species habitat affinity often is presented in tabular form as habitats used per species, or species present per habitat type, or in graphic form as maps or overlays.

The information collected in the literature search is compared with the description of the proposed alternatives for the project to determine the types of information required for evaluation of alternatives. If this process reveals gaps in the available information base, the presence of an endangered or threatened species, or the potential for significant impacts on wildlife or wildlife habitats, an optional field investigation may be undertaken. This investigation is designed to collect additional information on the presence, distribution, or abundance of a particular species; to determine the species of animals present in particular habitats and their relative abundance there; to determine the suitability of the various habitats for animals; or to identify the areas used for reproduction, feeding, watering, and shelter or concealment. A visit ordinarily is made to the project area to observe the proposed facility site. These observations are necessary to confirm the validity of the information obtained from the literature review and the applicant's data. Visual observations are recorded and photographic records prepared.

The third step in the assessment process is the identification and estimation of the primary and secondary impacts, both short- and long-term, of the applicant's proposal, and any feasible alternatives. In addition, information is needed on the location and acreage required for the facility from the applicant's plans and from any alternatives to the project.

Primary impacts are impacts that result directly from the construction and operation of the discharging facility, such as a loss of animal habitat resulting from the clearing of land for the system components, creation of artificial barriers to animal movement, and the effects on wildlife from noise, dust, odor, and other factors associated with the operation of construction equipment and increased vehicular traffic in the project area. Primary impacts are estimated by measurement of potential habitat loss from cover maps and by a written description of impacts on particular species or communities.

Secondary impacts are impacts that result indirectly from the implementation of the project, and usually are the effects of the induced growth (human population increase and changes in land use) stimulated by the increase in employment within the municipality or planning area. These impacts could include loss of animal habitat and other adverse effects on animals owing to human activity in the project area.

Identification and estimation of secondary impacts depend on the type of development stimulated by construction of the facility, which is described in the land use section of the EIS. The ultimate primary and secondary effects on animals often are similar. The estimation of impacts also includes identification of the potential benefits that result from the project, such as an increase in the suitability of habitat for certain species because of changes in the density or composition of plant cover, or the creation of protected buffer strips around the proposed facility.

The next step in the preparation of an EIS section dealing with wildlife is the estimation of the value of the existing project area habitats that may be affected by the proposed alternatives. The location and areal extent of these habitats are determined as part of the task on identification of sensitive areas. General descriptions of cover density usually are available from the vegetation survey. Areas with food sources required by particular species, areas that support economically valuable game species, or unique communities of plants and animals that may have high scientific, educational, or recreation values are described.

6.5 AQUATIC ORGANISMS

These organisms include:

- Algae (phytoplankton).
- Vascular aquatic plants.

- Zooplankton.
- Benthic organisms.
- Fish in various stages from larval to adult.

The areas to be studied include all water bodies in the project area or adjacent to it. Algae are sampled just beneath the surface or by taking several samples at vertical depths in the upper warm water layer. Sampling depends upon the body of water being investigated. Zooplankton samples are taken from different water levels. Vascular plants usually are mapped. Benthic organisms are sampled in the mud and silt at the bottom of the water body. Fish are captured either through gill nets or electroshocking. They are taken both upstream and downstream of the existing and proposed activities, as well as in nearby tributaries and other locations where spawning may occur.

In all of these cases, examination over a four season period is essential because of the wide variation of aquatic communities and life stages that can be expected to occur. An additional critical fifth season is often studied in case the original season was an unusual one.

Methods used for aquatic biological sampling comply with recommended and accepted methods as outlined in EPA's "Biological field and laboratory methods for measuring the quality of surface water and effluents" (1973) and the American Public Health Association's "Standard methods for the examination of water and wastewater" (1996). A detailed discussion of sampling methods follows.

6.5.1 ALGAE AND PHYTOPLANKTON

Phytoplankton, as primary producers, form the base of a complex food web by transforming radiant energy into foodstuffs available to higher trophic levels. The success of herbivores (plankton-feeding animals) depends on the condition and composition of the phytoplankton and other algal communities. Carnivore populations are supported indirectly by algae and other primary producers by obtaining food sources from herbivorous organisms. Phytoplankton are also important indicators of pollution, organic enrichment, and the trophic status in reservoirs.

Depending upon the objectives of a study, phytoplankton collections usually are made at several designated biological sampling stations. Whole water samples of approximately 2 liters are collected at each station. In a river that is well-mixed, only one sample at a 3 to 5 ft depth would be necessary; in a lake or reservoir, the mid-depth and near bottom both may be below the thermocline and thus 3 evenly spaced samples from the surface to the thermocline should be taken (Weber, 1973). The samples are immediately preserved. Each bottle is labeled with the sample type, date of collection, location, time of day, and collector and returned to the laboratory for analysis.

The numbers and types of the following major algal groups, as indicated by the American Public Health Association (1980) and Weber (1973), are determined:

- Greens.
- Blue-greens.

- Diatoms.
- Golden.
- Flagellates.
- Dinoflagellates.
- Others.

Phytoplankton identification and enumeration can be conducted with an inverted microscope and a phase contrast microscope. The Utermohl method of sample analysis described by Weber (1973) is utilized frequently. This method is chosen because the sample material receives a minimum of handling and the sampling and analytical protocol includes the nannoplankton. Counting procedures follow those outlined in Weber (1973). At least two strips (perpendicular to each other) across the bottom of the chamber are counted. The volume of water sedimented is adjusted to yield counts that will include at least 100 individuals of each of the most abundant species.

Phytoplankton density can be calculated and reported as numbers per ml for each species or major algal group. Relative abundance of each species and group can be calculated as the percent of the total algal density. Species diversity can be calculated using the Shannon-Weiner index (MacArthur and MacArthur, 1981) for each sampling station.

6.5.2 ZOOPLANKTON

Zooplankton have an important role in the food web of lentic environments by providing a trophic link between the primary producers and the macroinvertebrates, larval fish, and smaller forage fish species.

Zooplankton samples are collected at each of the sampling stations where phytoplankton were collected. Triplicate vertical (horizontal, if necessary) tows are made at each station, utilizing a number 20 mesh plankton net. The line used to lower and raise the net through the water column is marked at 0.5 m increments. The length of each tow is determined by the water depth at each specific sampling station. Samples are transferred to individual, labeled bottles and preserved.

The numbers, densities, and types of the following zooplankton groups then are determined microscopically:

- Copepods.
- Cladocerans.
- Rotifers.
- Others.

Relative abundances of each species and each major taxonomic group, and species diversity are determined as previously described for phytoplankton.

6.5.3 BENTHIC MACROINVERTEBRATES

Benthic organisms are an integral component of lentic environments because they serve as a principal source of food for various life stages of most fish species. In addition to their importance in the food chain, benthic organisms are sensitive to stress. The composition and characteristics of the benthic populations can serve as useful tools in detecting changes in the environment.

Because of their limited mobility and relatively long life span of a year or more, benthic macroinvertebrates may serve to indicate the recent past as well as present conditions. They also may indicate infrequently introduced stresses which may be difficult to detect by other means.

Benthic samples are collected from each designated sampling station, which should include representative areas of the profundal region, as well as littoral areas. Samples may be collected with any one of a number of benthic sampling devices including an Ekman dredge in soft areas or a Ponar sampler. Samples are washed through a standard No. 30 mesh sieve and the organisms and debris transferred to labeled containers and preserved. Organisms in each sample are handpicked or floated with a high density solution and are transferred to vials, and preserved.

The numbers and types of the following major benthos groups usually are determined:

- Aquatic oligochaetes.
- Aquatic chironomids.
- Other dipterans.
- Ephemeropterans.
- Tricopterans.
- Others.

Oligochaetes and chironomid larvae are mounted on microscope slides for identification and examined with a compound microscope. The microscopic mouth parts and teeth of chironomid larval are important features of identification. Other taxonomic groups are examined with an Olympus dissecting scope. Organisms are identified to the lowest practical taxon, using standard taxonomic references. Sampling, sorting, and subsampling procedures used often follow those described by Weber (1973).

Densities of each taxon and each major group are calculated as numbers per m^2 for each sample. Relative abundance (percent of total density), species diversity, and species richness are calculated for each station.

6.6 FISH

An accurate analysis of the fish population in a water body requires a sampling program that will account for all segments of the fish community. This may be accomplished by using the following collection methods.

6.6.1 ELECTROFISHING

Electrofishing, using a boat-mounted 230 volt alternate current generator, is conducted to sample fishes along the bank, shallow areas, vegetated areas, and other productive or protected areas. Shocked fish are dipnetted from the lake and placed in a live well. The number, size, and weight of individuals of each species are recorded. Scale samples for age determination may be secured. All fish are released unharmed as a result of the experience.

6.6.2   GILL NETTING

Experimental mesh monofilament gill nets are set at each of the sampling stations for approximately 24 hours. Each net consists of five panels of different size mesh. Additional individual nets may be set at the surface, thermocline, or near the bottom depending on the specific situation. Should aneroxic conditions exist at the bottom, the net should be set as close to the bottom as possible in oxygenated water. Fish are removed from each net and processed (weighed and measured). Following gill netting, there are few, if any, recoverable fish except for a few bullheads. Scale samples are taken as appropriate.

6.6.3   SEINING

Fish are collected with a haul seine in the shallow littoral areas. Seining will effectively sample young of the year centrarchids, cyprinids, and other small fishes. Seining should be conducted for a predetermined period at each station. Fish are processed after each seine haul, then released. Individuals not readily identified in the field are preserved and transported to the laboratory for positive identification.

The following parameters are determined for individual fish collected by each method:

- Species.
- Total length.
- Weight.
- Overall condition.
- Reproductive condition.
- Presence of disease and/or parasites.
- Age of, at least, representative size groups.

The combined efforts of electrofishing, gill netting, and seining will ensure the collection of a large number of individuals of each of the dominant species in the water body.

6.6.4   FISH TISSUE ANALYSIS

Fish tissue analysis is often performed to determine the presence of trace metals and chlorinated hydrocarbon pesticides. About five individuals of each species collected at each of the sampling stations are tested for trace metals. Two individuals of each species collected at each station are tested for pesticides.

A tissue sample (filet) is collected from each individual fish and wrapped in plastic bags (for metals analysis) or aluminum (for pesticides analysis), placed on ice, and transported to the laboratory. The objectives of some studies may require that whole fish be examined for bioaccumulative substances.

6.7 MITIGATION

Mitigative measures are available for some biological resource situations, while avoidance of negative impacts is the preferred approach for others. A brief summary of mitigative measures for the previous three sections follows.

6.7.1 TERRESTRIAL—UPLAND AND BOTTOMLAND COMMUNITIES

- Avoid areas where endangered and threatened plants occur.
- If possible, avoid wetlands. If not possible, minimize their use and create new wetlands to replace those used.
- Coastal zone areas, remnant and relict botanical areas—same as for wetlands.
- In other situations—minimize destruction or removal of plants. Landscape using native species to the extent possible.

6.7.2 WILDLIFE

- Avoid areas containing endangered or threatened species.
- Minimize habitat loss and noise, dusty vehicles, and so on during construction.
- In planning for the induced growth, attempt to include wildlife habitat areas.

6.7.3 AQUATIC ORGANISMS

- Avoid discharge of toxic contaminants into water bodies. NPDES and 404 permits should be used to control this, as well as the more conventional pollutants.
- Avoid destruction of spawning areas.

REFERENCES

MacArthur, R. and MacArthur, J. W., On birds species diversity, *Ecology,* 42, 594, 1981.

Revised guidelines for implementation of the Maryland Environmental Policy Act, Maryland Department of Natural Resources, June 15, 1974.

Standard methods for the examination of water and wastewater, 15th ed., American Public Health Association, American Water Works Association, and Water Pollution Control Federation, Washington, D.C., 1980, 1996. Weber, C. I.,

Biological field and laboratory methods for measuring the quality of surface waters and effluents, National Environmental Research Center, U.S. Environmental Protection Agency, Office of Research and Development, EPA-670/4/73-001, 1973.

7 The Man-Made Environment: Surface Water

The water-related environmental component of a NEPA study includes consideration of water quality, drainage patterns, nearby surface water bodies, and floodplains. Legal requirements for water quality must not be violated.

The initial portion of this chapter will review the legal requirements for water quality and the regulations that apply. The primary and secondary impacts on surface water will be defined. The methodology for defining the existing water quality then will be described. The evaluation of impacts will be discussed.

There are some very specific types of projects that may impact surface water. The EIS approach to some of these will be described in detail.

Requirements and impacts relating to groundwater are unique, and fit better in Chapter 8 of this book. They will be discussed there.

7.1 LEGAL REQUIREMENTS

The 1960s and 1970s brought forth a framework of federal laws providing for water quality protection. As a group, these laws were meant to protect human health, the water we drink, and the fish we consume. They were intended to protect aquatic life and to provide a quality suitable for recreation in and on the water. They are interrelated in that they are designed to function in concert, one with the other, in providing an umbrella of protection. Most have been amended since initially enacted in order to extend and solidify their initial requirements. While there are several laws affecting surface water that must be considered in the NEPA type studies, the two that are the most important by far are the Clean Water Act and the Safe Drinking Water Act. The provisions of those acts that must be considered in an EIS are discussed in the following pages.

7.2 THE CLEAN WATER ACT

7.2.1 WATER QUALITY STANDARDS

The key part of the Clean Water Act, insofar as NEPA is concerned, is the requirement to establish water quality standards. NEPA requires that any actions taken under it will not violate those standards.

The Clean Water Act calls upon the states to establish programs for water quality planning and management. The first part of this is the development of water quality standards. Each particular reach of each body of water in the state receives a set of goals for what the use of that water body should be. These goals could include any of the following: cold water fishing, warm water fishing, recreation, drinking water, aesthetics, and so on. The cleaner the body of water, the higher the goal for its use that the state would establish. Generally, water bodies tend to be at their cleanest in the mountains or other areas where they first form, and then they become polluted as they travel down towards the city. The water body tends to cleanse itself after it goes through the city, if additional pollution is not introduced, and eventually can be used for desirable purposes once more. For that reason, the water up in the mountains where the streams originate frequently will be designated for fishing for temperature- and oxygen-sensitive fish, such as trout. The next reach of water may be for recreational use, including both primary recreation (which means body immersion) and secondary recreation (which means boating). Water for drinking purposes may be taken from this reach of the river as well. As the water begins to approach the city and pass through it, the discharges from point sources and nonpoint sources become such that the utility of the water may be limited to boating and to appearance. Going through the city, the water may be limited to having aesthetic purposes. As the water goes past the city and begins to clean itself up, the higher uses begin to prevail. Eventually, the water body may empty into a lake or an ocean, and swimming and recreation may become the prime uses.

Having determined what the uses of a particular reach of water are to be, the state then decides what the chemical and physical criteria are for the water constituents that will affect those uses. For example, dissolved oxygen values of at least 6 ppm or higher are necessary for cold water fisheries. Generally, 5 ppm of dissolved oxygen is required for practically every use except the water that may serve an aesthetic purpose only.

Temperature is another sensitive indicator. For cold water fisheries, temperature requirements may be in the 60°F range. On the other hand, warm water fisheries may allow temperatures to go up into the 80°F range.

The bacterial count is particularly important in terms of the use of water for recreational purposes. The fecal coliform count is generally kept below 2 per cc so that swimmers do not get dysentery; 0.14 per cc protects shellfish harvesting. This means eventual restrictions on the treatment of sewage that might contribute fecal coliform.

Total dissolved solids are regulated, as is turbidity, because water clarity is a desirable item. In almost every case, grease, scum, and oil on the surface is forbidden.

Once these criteria for attaining the standards are set by the state, they must be approved as a part of the water quality standards by the administrator of the EPA. They then become the values that the particular water body must meet, as a minimum, to ensure its use for the designated purposes.

Another aspect of water quality standards is the nondegradation issue. That particular requirement was inserted many years ago to ensure that water bodies that have numerical values such as for temperature and dissolved oxygen that are much better than the minimum criteria for best uses, will not be degraded to the minimum levels without adequate consideration of the reasons for degradation and adequate public input. The author of this book was one of the authors of that nondegradation requirement that is now a part of every state's water quality standards.

Having established water quality standards, the states now are expected to establish and maintain a continuing water quality planning process that will ensure that those standards are met

The planning process may include such items as the following:

- Total daily maximum loads.
- Effluent limitations.
- Descriptions of best management practices for municipal and industrial waste treatment.
- Provisions for non-point sources.

How do industrial and municipal dischargers make certain that their discharge into water bodies will not upset the water quality standards requirements for their particular water bodies? The mechanism uses both effluent limitation guidelines and discharge permits.

As a result of several years of detailed studies of various types of industrial and municipal discharges, the EPA established effluent limitation guidelines for existing sources of water pollution, standards of performance for new sources, and pretreatment standards for certain types of both sources. These guidelines place limits on the quantities, rate, or concentrations of pollutants that may be discharged from point sources into a water body. They are based on what can be done for those discharges using the best available treatment technology. In addition, a list of 65 toxic pollutants has been published by the EPA that must not be discharged in toxic amounts into receiving water bodies. Limits are established on these toxic pollutants in the effluent guidelines. The state assures that these guidelines will not disturb the water quality requirement by doing very sophisticated water quality modeling on the body of water that will receive these discharges. Based on the modeling, determinations are made of how much in the way of contaminants can be introduced into a specific stretch of water without violating water quality standards. The state then determines how to best distribute the available quantities that may be discharged from the various point sources that have discharge requirements. The amount allocated to each source is written into permit requirements.

The situation is much more difficult in the case of nonpoint sources such as fertilizer runoff from farmlands or discharges from animal feedlots. Nevertheless, the state calculates how much in the way of pollutants from these sources may enter the water bodies, and what their effect will be on the water quality standards.

7.2.2 SECTION 404

Section 404 of the Clean Water Act is the mechanism for issuing permits for the discharge of dredged or fill material. It is the principal means within the Clean Water Act to prevent the unnecessary destruction of wetlands. Throughout its implementation, it has been a controversial part of the Act because the issues surrounding the granting or nongranting of a permit usually involve land development. Section 404 begins with four significant provisions; it states that

1. The U.S. Corps of Engineers may issue a permit, after notice and opportunity for public hearings, for the discharge of dredged or fill materials into the navigable waters "at specified disposal sites."
2. In specifying the disposal sites, the Corps of Engineers must use guidelines developed by the EPA in conjunction with the Corps.
3. Where the guidelines would prohibit the specification of a site, the Corps could issue a permit regardless, based upon the economic impact on navigation and anchorage.
4. The EPA is authorized to veto permitting a site based upon environmental considerations.

Regulations have been promulgated specifying how each of these actions will be managed.

7.2.3 NATIONAL POLLUTANT DISCHARGE ELIMINATION SYSTEM

The core of the Clean Water Act is the National Pollution Discharge Elimination System (NPDES), which requires anyone who discharges material into the navigable waters of the United States first to obtain a permit issued by the EPA or a state to whom permitting authority has been delegated. These permits limit the amount of pollution from each point source. The NPDES permit program operates in three stages: application, issuance, and compliance monitoring. Each stage involves a significant amount of information.

7.2.3.1 Application

Applicants must provide the permit-issuing agency with information on the production processes of their facilities, the characteristics of the effluents that result from these processes, and a description of the treatment methods they propose to use to control the pollution.

7.2.3.2 Issuance of NPDES Permits

The EPA Regional Administrator or responsible state official prepares a draft permit that consists of the appropriate effluent limitations for the point source, monitoring requirements, record-keeping requirements, and reporting obligations. It is then published for public comment, following which a final permit is issued. A discharge permit must not allow water quality standards to be violated.

7.2.3.3 Compliance with NPDES Permits

Individual permittees must provide information to the EPA or the state. The permittee must retain records that reflect all monitoring activities that are required in the permit. Monitoring and related activities must be conducted in accordance with the test procedures specified in the regulations. Discharge monitoring reports generally are required on a monthly basis.

7.2.4 OTHER KEY SECTIONS OF THE CLEAN WATER ACT

The preceding laws and regulations represent the key portions of the Clean Water Act with which most NEPA documents must conform. Construction grants, which until recently were perhaps the major federal activity that impacted NEPA, will be discussed later in this chapter.

Section 401 of the Clean Water Act is a significant section because it requires any applicant for a federal license or permit to obtain a certification from the state that any discharge connected with the action will not violate certain sections of the Clean Water Act, including existing water quality standards. No license or permit shall be granted if certification has been denied by the state, interstate agency, or the administrator of the EPA, as the case may be.

One other portion of the Clean Water Act that should be mentioned is the requirement for pretreatment of industrial discharges that flow to municipal waste treatment plants. These requirements are set by each local authority that operates the plants and conform to the EPA's pretreatment regulations. The purpose of the pretreatment program is to control pollutants that may pass through and interfere with the operations of the wastewater treatment plants or which may contaminate wastewater sludge.

7.3 THE SAFE DRINKING WATER ACT

7.3.1 STANDARDS

The Safe Drinking Water Act requires the promulgation by the EPA of primary drinking water regulations that specify maximum contaminant levels for constituents that may have any adverse effects on the health of persons, and of secondary drinking water regulations which specify maximum contaminant levels necessary to protect the public welfare. States have primary enforcement responsibility for the provisions of the Act, but must have EPA approval. Any NEPA activity that discharges into a supply of water to be used for drinking water purposes must keep this in mind.

The Safe Drinking Water Act contains a prohibition on the uses of lead pipes, solder, and flux in public water systems. EPA regulations place stringent limitations on the control of both lead and copper. The Act provides for the protection of underground sources of drinking water through the issuance of regulations for state underground injection programs, the provision of petitions by citizens for no new underground injection programs, and sole source aquifer protection where the vulnerability of an aquifer is owing to hydrogeologic characteristics. Amendments to the Act provide for a wellhead protection program and the identification of anthropogenic sources of contaminants to wells. Contaminant limitations promulgated under the Safe Water Drinking Act require filtration if the following contaminants do not meet EPA criteria:

- Total and fecal coliform.
- Turbidity.

Disinfection is required for most drinking water with a minimum of 0.2 milligrams per liter (1) of disinfectant residual maintained in the water entering the distribution system. The water must have the following degrees of inactivation:

- 99.9 percent of Giardia cysts.
- 99.99 percent of Enteric cysts.

The trihalomethane (THM) requirement in waters serving over 10,000 people is a total THM of less than 100 micrograms (μg) per l. A total coliform maximum containment goal of zero has been set.

The EPA's national primary drinking water regulations are found in 40 CFR Part 141. As of July 1, 1996, maximum contaminant levels had been set for the following chemicals:

- Subpart B, § 141.11 Inorganic chemicals—arsenic and nitrate.
- § 141.12 Organic chemicals—total trihalomethanes.
- § 141.13 Turbidity.
- § 141.15 Radioactive materials—radium-226, radium-228, and gas alpha particle activity.
- § 141.16 Radioactive materials—beta particle and photon radioactivity from man-made radionuclides.

Part 141 also contains a lengthy discussion of sampling and monitoring methods for a large number of chemicals. Part 142 lists maximum containment levels for many more organic and inorganic chemicals.

The National Drinking Water Advisory Council of the Environmental Protection Agency's Science Advisory Board, other federal agency officials, and the EPA have identified 58 chemical contaminants and 13 microbiological contaminants that may be targeted for future regulation, toxicity research, occurrence monitoring, or guidance development. The Safe Drinking Water Act (SDWA) Amendments of 1996 required the EPA to finalize a list of contaminant candidates by February 1998 and a monitoring list for no more than 30 of these by August 1999.

7.3.2 Drinking Water State Revolving Fund (DWSRF)

The material that follows is taken from the EPA program guidelines on DWSRF (EPA, 1997). The Safe Drinking Water Act (SDWA) Amendments of 1996 (Pub. L. 104-182) authorize a drinking water state revolving fund (DWSRF) to assist public water systems to finance the costs of infrastructure needed to achieve or maintain compliance with SDWA requirements and to protect the public health objectives of the Act. Section 1452 authorizes the EPA to award capitalization grants to states, which, in turn, can provide low cost loans and other types of assistance to eligible systems.

Under the SDWA, a state may administer its DWSRF in combination with other state loan funds, including the wastewater SRF, hereafter known as the Clean Water State Revolving Fund (CWSRF). Beginning one year after a DWSRF program receives its first capitalization grant (fund portion), a state may transfer up to a third of the amount of its subsequent DWSRF capitalization grant(s) to its CWSRF or an equivalent amount from its CWSRF capitalization grant to its DWSRF.

These two provisions linking the DWSRF and the CWSRF show congressional intent to implement and manage the two programs in a similar manner. The EPA will administer the two programs in a consistent manner and will apply the principles developed for the existing CWSRF to the DWSRF program. Each state will have considerable flexibility in determining the design of its program and in directing funding toward its most pressing compliance and public health protection needs. Only minimal federal requirements will be imposed.

The DWSRF has been authorized at $9.6 billion over a 10 year period ending in Fiscal Year 2003. The EPA began awarding state capitalization grants in early 1997.

7.4   MAJOR WATER POLLUTION PROJECTS
SUBJECT TO NEPA

In this section, we will review two types of water pollution projects that are subject to NEPA:

- Municipal wastewater treatment plants.
- New sources that require NPDES permits.

A brief discussion of each type of project and the NEPA elements involved will be presented. These types of projects have accounted for the majority of EPA's water-related NEPA compliance activities.

7.4.1   THE MUNICIPAL WASTEWATER TREATMENT PLANT
PROGRAM

NEPA compliance procedures apply to all municipal wastewater treatment plant construction grants projects that received Step 1 grant assistance on or before December 29, 1981, approval of grant assistance for a project involving Step 3 or Steps 2 and 3; and an award of grant assistance for a project with significant changes in the scope or impact of the project. The step designations relate to the state of planning and design. The environmental review procedure followed in implementing NEPA compliance requirements includes five steps. For all practical purposes, the construction grant program ended in 1990. However, this material is shown here because even after all this time, some projects still are in these steps.

1. The first step in the process is *consultation.* The principal activity included in the consultation process is to determine whether a project is eligible for a categorical exclusion from the remaining steps in the environmental review process. Other key points to address here include identification of possible alternatives, identification of potential environmental issues, opportunities for public recreation and open space to be developed as part of the project, the potential need for partitioning of the project, and an early consideration of the potential for the need of an EIS.

2. The second step in the NEPA compliance process is the actual determination of the project's *eligibility for a categorical exclusion.* The potential for environmental impacts resulting from wastewater construction grants

projects was diminished substantially by the changes in the program resulting from the 1981 Amendments, which prohibited granting of funds for development of facilities to serve the future population. This is owing to the fact that much of the environmental impact of sewer facilities comes from indirect impacts caused by population growth and land development supported by the facilities. Based on the regulations promulgated in response to the Construction Grants Amendments of 1981, it is estimated that as much as 20 percent of the EPA-funded projects were excluded from substantial environmental review. Some of the types of construction grants projects which may be eligible for categorical exceptions include:

- Minor rehabilitation of existing facilities.
- Functional replacement of equipment.
- Construction of new ancillary facilities.
- Minor upgrading and minor expansion of existing treatment works in unsewered communities of less than 10,000 persons.

3. The third step in the compliance process is *documenting environmental information.* The Environmental Impact Document (EID) must include all of the general environmental information about the proposed facility. One of the specific issues related to the development of a facility plan EID is the need to provide sufficient detail to enable a decision on partitioning. Partitioning refers to the identification of clear phases of the facility plan, so that certain components can be constructed in advance of completing NEPA requirements for remaining portions of the project. The criteria utilized in making a determination on partitioning for a component include:

- The component's use as an immediate remedy to a severe public health, water quality, or other environmental problem.
- It must not foreclose reasonable alternatives for the overall system.
- It must not cause significant adverse direct or indirect environmental impacts.
- The component also must not be highly controversial.

4. The fourth step in the NEPA compliance process for facility grant projects is *preparing environmental assessments.* This phase of the work includes the preparation of an EA by the EPA, or, in the case of a delegated state, the state prepares a preliminary EA for review and approval by the EPA. Based on the results of the EA, either a finding of no significant impact (FONSI) or a notice of intent to do an EIS is prepared. The specific criteria used in making an EIS determination for a construction grants project include assessing whether:

- The facilities (including sludge management system) will induce significant changes in land use.
- The treatment works, including the collection system, will have significant adverse direct or indirect effects on wetlands.
- There is potential for significant adverse impacts on threatened or endangered species.
- The potential exists for direct or induced changes in population.

- Adverse effects may result on floodplains, parklands, public lands, and areas of recognized scenic, recreational, archaeological, or historic value.
- There may be significant adverse direct or indirect effects on local ambient air quality or noise levels.
- The treated effluent will continue to be discharged into a body of water for which the present classification is too low to protect its use.
- The treated effluent will have a significant adverse impact on existing or potential sources of groundwater supply.

In making this determination, the responsible official also must consider whether the project is highly controversial; whether it may produce significant cumulative impacts; or if the proposed facilities would be in violation of any other environmental law. When the decision to prepare an EIS is made, the procedure followed will be basically be the same as outlined above.

Following issuance of the final EIS, the responsible official issues a record of decision (ROD). The ROD must include identification of mitigation measures derived from the EIS process including grant conditions necessary to mitigate adverse impacts of the selected alternative.

5. The final step in the compliance process is *monitoring*. Monitoring of construction grants projects for compliance with EIS results includes construction and post-construction operation and maintenance of the facilities and review of compliance with any grant conditions.

As a result of changes deriving from the Water Quality Act of 1987, the construction grants program has been replaced by a state revolving fund (SRF) as a source of funding for municipal wastewater treatment plant construction. The SRF is a much broader program than the construction grants program in that its funds may be used for financing a wide variety of environmental infrastructure projects, for example, wastewater treatment, agricultural and urban runoff, stormwater, combined sewer overflows, excess capacity, collection systems, and so on.

Funds for the SRF program are provided through Federal grants (83 percent) and state matching funds (17 percent). As of 1995, these funds totaled more than $16 billion. All 50 states and Puerto Rico were operating successful SRFs at that time.

The basics of the SRF program are very simple. The federal and state contributions are placed in the fund. Communities borrow the money at interest rates of anywhere from 0 percent to market rates. The repayment (which begins one year after project start-up) is up to 20 years, with provisions for earlier payments at state discretion.

On June 19, 1998, the EPA Assistant Administrator for Water made public the results of a survey by the EPA on the results of the SRF to date. He concluded that states are not leveraging as much money as they can from loans to construct or upgrade wastewater treatment plants and should expand their programs to include more nonpoint source pollution control projects and to improve water quality in estuaries.

NEPA studies on the activities that are funded through the SRF program are required or not required by the same set of standards used for other federal activities. It must be determined whether or not the action is a major federal action. The normal NEPA procedures then are followed.

7.4.2   NEW SOURCE NPDES PROJECTS

Approval of NPDES permits for new source discharges is another major program area where EPA has direct NEPA compliance responsibilities. All potential point source discharges must obtain a permit to discharge from either the EPA or a state agency authorized to administer the NPDES. For industrial dischargers, these effluents are subject to new source performance standards (NSPS) which are promulgated by the EPA for specific categories of industries.

The first step in the NEPA compliance procedures for new source NPDES projects is to determine the applicability of NEPA to a specific permit application. Based on current EPA regulations, NEPA requirements apply only to the issuance of discharge permits to *new sources* located in states that do not have approved state NPDES permit programs. Further, only certain categories of projects are defined as new sources. The EPA defines a new source (40 CFR 122) as a building, structure, facility, or installation from which there is or may be a discharge of pollutants and on which construction was begun after new source performance standards applicable to the source were proposed. New construction can include a totally new source, modification of an existing source, or a major alteration to an existing source. Modifications to an existing source which already has a discharge permit, including changes in production capacity by adding a process unit to the existing facility, are not considered new sources and are subject only to permit modification procedures. Existing sources can be defined as new sources if major alterations are involved. A major alteration would include the construction of an additional facility or facilities on the existing site which function independently of the existing discharge.

Once it is determined that NEPA requirements apply, the remaining steps in the environmental review procedure for new source NPDES permits are undertaken. The basic steps in this process generally are similar to those described previously for construction grants projects—EID, EA, FONSI, or EIS, ROD, and monitoring. The critical issues and differences associated with this program have to do with the identification of a lead agency and application of the criteria for preparation of an EIS.

Unlike construction grants projects, where the EPA is in most cases the federal agency with primary authority and responsibility for the action under review, a number of different agencies also may have major involvement in new source NPDES projects. To make the determination of lead agency, the responsible official must contact all other involved agencies and together decide the lead agency, using criteria established by the Council on Environmental Quality (40 CFR 1501.5). The factors to be considered in determining a lead agency are:

- Magnitude of agency involvement.
- Project approval or disapproval.
- Expertise concerning the environmental effects.

- Duration of the agency's involvement.
- Sequence of the agency's involvement.

7.5 ENVIRONMENTAL IMPACTS

7.5.1 THE AFFECTED ENVIRONMENT

A detailed description of the existing surface water environment must be presented. This thorough assessment involves a detailed inventory and characterization of surface water resources in the project region, complete with quantity and/or quality relationships, and an identification of regulatory standards and water quality classifications applicable to local surface waters. Surface water-related problems in the project area are clearly identified, and current baseline information is established to enable an accurate assessment of impacts. Potable water supplies and systems, as well as wastewater treatment plant effluent contributions and various point source and nonpoint source contributions, also are included in this assessment. Water flow, drainage patterns, and floodplains must be described.

In many situations, all of the factors required to establish the baseline conditions may not be known. In those cases, field sampling may be required. Available information and the objectives of the study are used to develop the sampling methodology. Field investigations may range from one-time reconnaissance surveys to year-long inventories of physical, chemical, and biological characterizations. The coordination of chemical and biological sampling is necessary to assure accurate interpretation of data for the characterization of existing conditions and the evaluation of primary and secondary impacts. A detailed discussion of sampling techniques is found in *Environmental Regulations,* Chapters 6 and 7, by K. M. Mackenthun and J. I. Bregman.

7.5.2 IMPACT METHODOLOGIES

The effects of projects subject to NEPA requirements on surface water resources of a region often are complicated and should be approached systematically. A flexible approach or methodology should be used rather than a rigidly structured system to allow for variation from project to project.

Primary impacts on water resources are directly related to the construction and/or operation of the proposed project. Impacts encountered most frequently are water quality degradation or improvement resulting from operation of a proposed facility, possible siltation of nearby waters during construction, and increased or decreased streamflow from addition or reduction of waste discharges.

Secondary impacts on water resources are those related to growth and development. Population growth and associated land-use changes affect water quality by altering part of the natural hydrologic cycle—precipitation, infiltration, surface and subsurface runoff, and stream flow. Water quality also is affected by the addition of pollutants during one or more parts of this cycle and by discharges from man-made facilities. Exhibit 3 illustrates significant interrelationships between types of

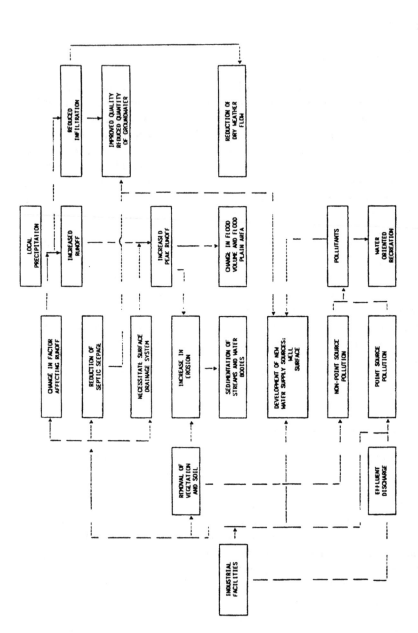

Exhibit 3 Interrelationships between industrialization and primary and secondary impacts on surface waters. (Modified from *Manual for Evaluating Secondary Impacts of Wastewater Treatment Facilities*, U.S. Environmental Protection Agency, 1978, 35. With permission.)

environmental changes resulting from development and their subsequent impacts on water quality and quantity. While the interrelationships are complex, key variables for determining secondary impacts of industrialization include: increased total and peak run-off volumes, erosion and sedimentation, increased point and nonpoint pollutant loadings, and quality and quantity of water supplies.

7.5.3 Water Flow

Growth and development induced by construction or expansion of wastewater or industrial facilities may increase sewage flows. Additional flow then will be conveyed to treatment plants, discharging more flow to surface waters. In addition, the amount of urban runoff will increase because of impervious surfaces and less infiltration. These additional flows may contribute larger pollutant loads to the water and may have an adverse effect on water quality.

To determine primary and secondary impacts of municipal or industrial wastewater facilities on flow, it is necessary to characterize existing water flow in the project area. Stream-flow records are analyzed by determining historical trends and seasonal variations and by comparing yearly and seasonal low flows with 7 day, 10 year flows.

Necessary data usually can be obtained from the U.S. Geological Survey (USGS) records and from EPA's Water Quality Control Information System (STORET). STORET contains information from federal and state sources, although information for a specific stream segment may not be available. In these cases, information from the nearest gaging station is used, or flows are estimated by using data from the nearest gaging stations and by adjusting values based on larger or smaller watershed areas. Other potential sources include state geological surveys, the U.S. Army Corps of Engineers, Section 208 water quality planning agencies, and USGS hydrologic investigation atlases. If it cannot be found in any of these locations, it must be measured.

Changes in flow can affect water quality both upstream and downstream from the project. If upstream flow is reduced, the stream assimilates smaller pollutant loads from point and nonpoint sources. If flows discharged upstream contribute significant loads, the water quality will improve when the flows are reduced or eliminated. The stream's assimilative capacity (downstream from the treatment facilities) may be reduced significantly if pollutant loads are discharged at one point.

Flows also must be measured to assess possible impacts on the volume of the receiving body of water. Data on flows conveyed to a treatment plant often can be obtained from plant records. Domestic flows discharged to on-site systems can be estimated by determining the area's residential water consumption rate and/or by estimating the number of households in the area, and then by assuming a wastewater discharge rate. Industrial flows can be determined from plant records, process water consumption rates, and self-reporting permit forms.

After existing wastewater flows in the project area have been estimated, wastewater management alternatives are reviewed to determine changes in wastewater flows. Flows resulting from induced growth will be the difference between flows estimated for the no-action alternative (using the population projections based on no-action) and flows estimated for the other alternatives (using the population projections based on the construction of new or additional facilities). Impacts of changes on streamflow can be evaluated qualitatively. If impacts appear to be potentially significant, a quantitative analysis employing a hydrologic model can be conducted for further evaluation.

7.5.4 WATER QUALITY

An accurate evaluation of water quality impacts depends on good characterization of existing conditions and sound predictive techniques. Information on the present water quality of a study area, including existing water quality problems, is usually available from the USGS WATSTORE system, the EPA STORET system, or state and local water regulatory agencies.

After collecting and reviewing available data on organic, nutrient, bacterial, and solids loadings, critical data gaps are filled, where necessary, by verification spot checks on streamwater quality.

Upon completion of the baseline water quality inventory, environmental constraints (both physical and regulatory) that must be considered in the impact analysis of the alternatives should be determined. Physical constraints may include limited water supply, poor soils, high groundwater tables, and nonpoint source pollution, or a mixture of agricultural and urban nonpoint sources. Typical regulatory constraints include water quality standards and effluent limitations.

With background data, information on water quality constraints, and a knowledge of estimated wastewater characteristics for point or nonpoint sources, a variety of predictive tools may be used in assessing water quality impacts. A preliminary assessment first is made to determine the relative magnitude of the pollution source being examined. Nonpoint sources, for example, may be of such magnitude in an area as to mask any benefits that may be derived from improved treatment of a point source.

The principal beneficial impact of wastewater management alternatives is improved water quality resulting from the reduction of pollutant loads to surface waters. Negative primary impacts on water quality can result from construction, expansion, rehabilitation, and upgrading of municipal and industrial wastewater facilities. Construction activities can increase sediment loads to surface waters and cause short- and/or long-term impacts. In addition, a new point source discharge (from a new industrial or municipal treatment facility) or a discharge contributing larger pollutant loads at one point (from expanded, rehabilitated, and/or upgraded facilities) can have adverse water quality impacts in and beyond the mixing zone (downstream if the discharge is to a river or a stream).

Secondary impacts on surface water quality can result from construction of new facilities or the expansion of existing facilities, both of which can induce growth and development. Construction activities may increase sediment and nutrient loads to surface waters. In addition, municipal or industrial development can degrade runoff quality because of increased pollutant loads of sediment, organics, bacteria, and heavy metals.

Computer models can be used to estimate pollutant loads (e.g., sediment and nutrient loads) discharged from point and nonpoint sources. The characteristics of the point and nonpoint source discharges are then projected based on wastewater management alternatives and future socioeconomic conditions (e.g., projected populations and land use).

By using information on present and future discharges, water quality impacts can be evaluated qualitatively through comparing present pollutant loads with projected pollutant loads. To evaluate impacts quantitatively, present concentrations of relevant water quality parameters can be compared with concentrations based on the loading conditions of various alternatives. Present water quality concentrations can be obtained from available sources of information and field investigations, or estimated by computer modeling.

Estimating concentrations by modeling existing and future conditions requires information on background concentrations, pollutant loads, and characteristics of the receiving water body.

In many cases, surface water quality changes can be predicted in terms of dissolved oxygen concentration, temperature, dissolved solids concentrations, and/or nutrient concentration, using a mathematical computer model.

Possible modeling applications include:

- NPDES permit evaluations.
- Stream assimilative capacity analyses.
- Waste load allocation studies.
- Nonpoint source pollution evaluations.
- Stormwater modeling studies.
- Waste heat disposal/thermal pollution analyses.
- Eutrophication analyses.
- Ocean outfall/waste-disposal-at-sea studies.
- Sediment transport analyses of stream and estuarine systems.
- Toxic substance modeling studies.

A wide variety of on-line models are available. Some of them follow:

- DOSAG-I (DO, BOD).
- QUAL-II (BOD, DO, temperature, NH_3, NO_3, NO_2, algae, phosphorus, benthic demand, coliforms, radioactive materials, three conservative constituents).
- STORM (stormwater runoff).
- SWMM (stormwater collection, treatment, storage, and water quality).
- HEC (streamwater profile and sediment transport).
- WRE deep reservoir model (far-field temperature prediction).
- USGS groundwater model (2-D aquifer simulation).

7.5.5 WATER SUPPLY

The construction of new facilities or the expansion of existing facilities can have primary and secondary impacts on water supply. Primary impacts on water supply can result from construction and operation. Construction can increase erosion and sediment load to surface supplies. Operation of new or expanded facilities can discharge an effluent that may have an adverse effect on the water quality of downstream supplies. Any significant impact on the water quality can increase the cost of water purification.

Impacts of increased pollutant loads on water supply can be determined qualitatively or quantitatively. Background water quality and existing and projected loads determine the impacts on a downstream water supply. Both pollutant loads and the distance between the plant outfall and the supply must be evaluated. Stream modeling can estimate impacts on downstream supply by using data on background concentrations and pollutant loads and time-of-travel studies.

Secondary impacts on water supply can result from growth induced by new construction. Development can increase sediment and other pollutant loads to surface supplies (via run-off), and thus can affect water quality adversely. More importantly, induced growth and development may increase water demands to exceed available supply, or may cause water shortage problems during extended dry-weather periods.

Drinking water models and databases include the follwing:

> *EPANET*—a computer program that performs extended period simulation of hydraulic and water quality behavior within drinking water distribution systems.
>
> *National Contaminant Occurrence Database*—NCOD is a new database being developed to help the EPA track contaminants in drinking water.
>
> *Pollutant Routing Model, Windows (P-ROUTE)*—a simple routing model that estimates aqueous pollutant concentrations on a reach by reach flow basis, using 7Q10 or mean flow.
>
> *Safe Drinking Water Information System (SDWIS/FED)*—the EPA's national regulatory database for the drinking water program, available through *Envirofacts*.

7.6   FLOODPLAINS

Floodplains are defined as lowlands and relatively flat areas adjoining inland and coastal waters, and include, at a minimum, areas subject to a 1 percent or greater chance of flooding in any given year.

Executive Order 11988, Floodplain Management (May 25, 1977), requires that federal agencies evaluate the potential effects of their actions in floodplains to avoid adverse effects associated with their direct or indirect development. Complying with this Executive Order requires a determination that no practicable alternative exists to proposed actions and that all appropriate mitigating measures are applied.

The existence of floodplains on or near the project area should always be documented in the section of the EIS that deals with the existing environment. The nature of floodplains, their extent, and the presence of any man-made facilities, roads or other activities on them should be noted. The extent of the 100-year flood hazard from nearby water bodies that might come on or close to the project property should be noted.

In general, most localities and states discourage or prohibit man-made facilities in a floodplain for the following reasons:

- The facility to be created is subject to possible damage or destruction by floods.

- Insurance coverage is move expensive and more difficult to obtain.
- The paving over of part of a natural floodplain will serve to extend the flow of water into other directions with possible severe impacts upon man-made facilities that already exist there.

Thus, a project may not only have an undesirable impact on a floodplain, but the floodplain might also cause a disaster to the project at some future date.

Any likelihood of induced development in a floodplain should be determined by analyzing future growth coupled with land-use control measures (local ordinances and Flood Insurance Administration programs). Induced growth also may contribute to increased flood levels by increasing the amount of impervious area within a watershed. That, in return, may decrease the time of concentration and increase both the peak and volume of run-off.

The floodplain problem is not an academic one. All too often, industries have been placed in floodplains because of their need for ready access to large nearby rivers for water for industrial purposes, such as cooling waters. In addition, housing developments and recreational facilities have been located in floodplains because the land is relatively inexpensive compared to nearby higher ground. It thus becomes critical to obtain floodplain information early in the EIS process so that the applicant may be made aware, as rapidly as possible, of the danger to his project.

Information on floodplains is most conveniently obtained by examining Flood Insurance Rate Maps prepared by the Federal Emergency Management Agency (FEMA). These maps are not always available, especially for remote or sparsely populated areas. These large-scale maps identify the boundaries of a 100 year flood hazard, and sometimes identify the boundaries of a 500 year flood hazard. These hazard areas are shown as darkened areas that include the appropriate stream or water body. An investigator is forewarned that identifying landmarks on these maps are minimal, so it is essential to be quite familiar with the geography of the area of interest before the maps are examined.

Obtaining Flood Insurance Rate Maps from FEMA generally is a two-step process. There is a toll-free telephone number, but the maps are sold by particular identifying panel numbers. First, it is necessary to obtain the index map for the county in which a potential project is located. The county index map is divided into panels with the identifying number clearly indicated as an aid to ordering the particular panel or panels of interest. Some areas may not subscribe to the FEMA program and, therefore, a Flood Insurance Rate Map will not have been produced. The U.S. Geological Survey (USGS) has delineated flood-prone areas at the 1 : 24,000 scale for many 7.5 minute topographic quadrangles. These data are also useful when they are available. The U.S. Army Corps of Engineers has determined flood elevations for some major navigable waterways. These data can be used where FEMA and USGS data are lacking. Local water resource agencies, state water agencies, or geological surveys may have useful information.

If a project involves construction activities, floodplains are areas to be avoided. Generally, it is prudent to locate the construction activities outside of a 100 year floodplain area. Wetlands are usually closely associated with floodplains. Where wetlands occur, a Clean Water Act Section 404 permit would be required for construction activities. In addition, a Clean Water Act Section 401 State Certification most probably would be required. If the potential project is intended to be located in a coastal zone, there is a further required certification that the applicable coastal zone management plan would not be significantly impacted.

REFERENCES

Mackenthun, K. M. and Bregman, J. I., *Environmental Regulations Handbook,* Lewis Publishers, Boca Raton, FL, 1992.

Drinking water state revolving fund program guidelines, U.S. Environmental Protection Agency, Office of Water, EPA 816-R-97-005, February 1997.

Manual for Evaluating Secondary Impacts of Wastewater Treatment Facilities, U.S. Environmental Protection Agency, 1978, 35.

The clean water state revolving fund, U.S. Environmental Protection Agency, Office of Water, EPA 832-R-95-001, January 1995.

8 The Man-Made Environment: Groundwater

In recent years, the need to protect groundwater has become a critical part of all new construction projects. This increasing emphasis is because of the rapidly growing use of groundwater as a source of drinking water. According to Briggs (1976), more than 97 percent of the world's fluid fresh water is underground. Obviously, this amount of water is not all available as a source of future drinking water, but its magnitude defines underground water as a precious resource to be husbanded for future needs. The use of groundwater for all purposes in the United States was estimated by the U.S. Geological Survey (Salley, 1997) to be as shown in Exhibit 4 for the year 1995.

In contrast to rapidly moving freshwater streams, underground water moves very slowly and does not have ready access to oxygen supplies. This means that contaminated underground water tends to remain in that condition. The result all too often is the loss of a major source of water supply for public or industrial purposes.

8.1 REGULATORY BACKGROUND

Recognizing the need to keep underground water from becoming contaminated, the U.S. Congress, the EPA, and a number of states have taken various protective actions. There are a number of different federal and state laws and regulations that are designed to protect groundwater. In contrast to other media, groundwater protection is a function of several laws.

The most common ones place emphasis on three areas:

- Contamination by buried hazardous wastes.
- Maintenance of the purity of sole-source aquifers.
- Underground injection of wastes.

The first, contamination of groundwater by buried hazardous wastes, is regulated by RCRA, CERCLA, and the Safe Drinking Water Act, which are discussed elsewhere in this book. Of particular concern is the possible contamination of groundwater by leachates from landfills. Existing landfills are covered by RCRA and abandoned ones by CERCLA. Over the years, toxic chemicals have accumulated in landfills and are leaching into the nation's groundwater supply. Existing landfills now

EXHIBIT 4
Preliminary Groundwater Withdrawals by Water-Use Category and State, 1995*

State	Public Supply Fresh	Domestic Fresh	Commercial Fresh	Irrigation Fresh	Livestock Fresh	Industrial Fresh	Industrial Saline	Mining Fresh	Mining Saline	Thermo-Electric Fresh	Total Fresh	Total Saline
Alabama · · · · · · · ·	253	62	4.9	51	22	34	0	4.0	9.1	6.0	436	9.1
Alaska · · · · ·	30	8.3	11	0.1	0.1	3.8	0	0	75	4.2	58	75
Arizona · · · · ·	409	39	21	2,130	29	39	0	119	12	42	2,830	12
Arkansas · · · · ·	135	38	0.4	4,930	244	108	0	0	0	5.2	5,460	0
California · · · ·	2,740	108	75	10,800	234	522	10	14	151	3.6	14,500	183
Colorado · · · · ·	100	27	7.7	2,020	23	37	0	25	17	22	2,260	17
Connecticut · · ·	65	55	25	16	1.4	3.5	0	0.3	0	0.2	165	0
Delaware · · · · ·	40	12	2.8	34	3.8	17	0	0	0	0.2	110	0
District of Columbia	0	0	0	0	0	0.5	0	0	0	0	0.5	0
Florida · · · · ·	1,860	297	50	1,670	50	240	0	148	0	21	4,340	4.6
Georgia · · · · ·	263	99	36	476	9.7	295	0	8.7	0	4.8	1,190	0
Hawaii · · · · ·	200	2.4	45	173	7.5	19	0.9	0.5	0	67	515	16
Idaho · · · · · ·	180	65	9.8	2,520	17	39	0	1.2	0	0	2,830	0
Illinois · · · · ·	371	129	16	180	54	162	0	5.5	25	11	928	25
Indiana · · · · ·	319	115	45	61	28	119	0	10	0	11	709	0
Iowa · · · · · ·	257	45	18	35	82	74	0	1.1	0	15	528	0
Kansas · · · · ·	181	24	4.9	3,150	91	35	15	13	0	14	3,500	15
Kentucky · · · ·	55	23	8.0	0.5	2.3	92	0	7.4	0	38	226	0
Louisiana · · · ·	294	39	10	475	144	356	0	0.4	0	31	1,350	0
Maine · · · · · ·	25	35	9.5	2.6	1.4	4.6	0	1.3	0	0.7	80	0
Maryland · · · ·	83	73	19	37	13	19	0	0.9	0	1.8	246	0
Massachusetts · · ·	192	34	12	28	1.5	38	0	0.5	0	46	351	0
Michigan · · · · ·	348	194	16	101	13	177	3.6	7.1	0.8	3.6	858	4.4
Minnesota · · · ·	331	88	46	120	62	57	0	6.3	0	1.9	712	0

Mississippi	302	33	16	1,540	377	166	0	3.5	0	42	2,590	0
Missouri	226	58	13	535	20	27	0	8.6	0	9.5	891	0
Montana	55	17	0	82	16	31	0	2.6	13	0	204	13
Nebraska	232	42	0.3	5,780	108	26	0	6.1	4.7	4.4	8,200	4.7
Nevada	117	11	7.1	641	1.0	47.4	0	65	11	6.3	855	42
New Hampshire	31	31	12	0.3	0.6	5.6	0	0	0	0.8	81	0
New Jersey	387	88	17	32	1.5	43	0	2.4	0	1.9	580	0
New Mexico	277	28	18	1,280	26	6.3	0	81	0	9.3	1,700	0
New York	552	144	136	16	22	127	0	11	1.5	0	1,010	1.5
North Carolina	134	172	7.3	57	69	61	0	12	0	0.1	333	2.1
North Dakota	30	12	0.1	59	14	3.6	0	3.8	0	0.3	122	0
Ohio	497	138	28	12	7.6	158	0	47	0	19	905	0
Oklahoma	99	30	6.6	755	45	3.8	0	5.4	259	3.5	959	259
Oregon	124	61	4.4	880	3.1	13	0	1.2	0	0	1,090	0
Pennsylvania	243	145	16	5.2	48	85	0	211	0	6.2	762	0
Rhode Island	16	7.3	1.5	0.7	0.5	1.1	0	0.5	0	0	27	0
South Carolina	107	71	1.7	27	12	80	0	2.9	0	39	322	0
South Dakota	53	9.3	6.1	65	18	4.1	0	7.8	0	3.4	167	0
Tennessee	277	54	2.0	9.9	21	66	0	2.8	0	0	435	0
Texas	1,130	130	33	6,530	139	228	0.5	128	409	59	8,370	411
Utah	293	7.7	3.8	393	7.8	55	0.1	16	7.3	0	776	14
Vermont	15	18	9.5	0.4	4.6	1.9	0	0.3	0	0.4	50	0
Virginia	82	125	28	5.8	7.8	107	0	2.6	0	0.4	358	0
Washington	631	125	24	819	24	133	0	2.8	0	0.5	1,780	0
West Virginia	38	40	38	0	15	13	0	3.7	0.5	0.5	146	0.5
Wisconsin	311	92	17	167	79	78	0	7.9	0	5.8	759	0
Wyoming	38	9.7	0.9	181	13	1.6	0	71	18	1.0	317	18
Puerto Rico	95	6.4	1.2	33	4.5	10	0	2.8	0	2.2	155	0
Virgin Islands	0.3	0	0.1	0	0.1	0.1	0.2	0	0	0	0.5	0.2
Total	15,100	3,310	940	49,000	2,280	4,010	30	1,060	1,010	565	76,300	1,130

*Figures may not add to totals because of independent rounding. All values are in million gallons per day.

are stringently regulated insofar as what toxic materials may be placed in them. Liquids are banned. Furthermore, double liners usually are required around the landfills in order to trap liquids leaching from the wastes. These liquids then are pumped up into suitable containers and disposed of properly. Finally, only a very few landfills are allowed to handle toxic materials, perhaps as few as one or two per state. Even the number of conventional landfills has dropped sharply as many permits have expired and were not renewed.

In order to obtain permits under RCRA, owners and operators of hazardous waste facilities usually must install groundwater monitoring wells to detect contamination from the facilities and undertake the necessary corrective measures.

The EPA has adopted a twofold approach to RCRA permitting standards:

- Liquids management.
- Groundwater monitoring and response.

The first, liquids management, minimizes the generation of leachate that might contaminate groundwater. The second compares groundwater contamination to that which occurs naturally in the uppermost aquifer or the maximum contaminant levels (MCLs) for contaminants listed under the Safe Drinking Water Act, whichever is less. If those values are exceeded, then corrective actions must be taken.

Sole-source aquifers are those which are the sources of present or future water supply to an area where there is only one aquifer available for that purpose. Consequently, these aquifers are subject to rigid regulations that are intended to protect them from becoming contaminated. The Safe Drinking Water Act calls on states and local governments to identify critical aquifer protection areas that are eligible for special protection. EPA regulations provide criteria for states and local governments to use for the identification of such aquifers. This necessity is especially true in states like Florida and New Jersey which face precarious future water supply situations.

In doing a NEPA study, therefore, any possible effect of the proposed project on sole-source or critical aquifers must be given close scrutiny. As a corollary to this, aquifer recharge areas and the effects of the proposed project on them also must be considered.

For many years, the underground injection of wastes, for example, cyanide and heavy metals from the steel industry, was a commonly accepted practice. These wastes were injected into aquifers that thus became contaminated and therefore unsuitable for future use as public water supplies. In recent years, the EPA has placed increasingly stringent restrictions on this practice and has reduced the amounts of permitted underground injection activity drastically. The apparent goal is eventually to eliminate this practice entirely.

For those applicants who desire to start or continue underground injection, the EPA has placed underground injection wells in five categories or classes according to the nature of the substance and the threat it poses. The EPA considers whether or not to allow operation of those wells through one of two mechanisms—by general rule or by individual permit.

In the cases of those underground injection wells that the EPA allows to function *by general rule,* the owner/operator is required to submit to the EPA or the state that has been delegated authority an inventory of the underground injection wells and, for certain classes, the rates of injection. Reports also may be required on groundwater monitoring and analyses of fluids injected into the wells. Thus, for Class I wells the owner/operator is required to:

- Install and maintain groundwater monitoring wells.
- Monitor continuously injection pressure, flow rate volume, and other characteristics.
- Analyze injected fluids.
- Report quarterly data on injection pressure flow rate, volume of injected fluids, results of groundwater monitoring, and any tests of the well ordered by the agency during the period.

When authorized *by permit,* the EPA or the state must require a form of monitoring that will produce data representative of the activity being monitored. The permittee is required to present these results (40 CFR 146). The permit will also require the owner/operator to report any changes in the facility. Any incident of noncompliance with permit conditions that endangers an underground source of drinking water must be reported to the agency within 24 hours.

These regulatory controls ensure the mechanical integrity of the injection well operations, as well as ensure that none of the well's contents migrate into underground sources of drinking water. Information collected subsequent to authorization by rule or permit enables the EPA or the states to remain informed about the contents and operating characteristics of these wells and the status of groundwater in the area. In this way, changes in rules or permits can be made and, when necessary, enforcement actions can be taken.

Leaking underground storage tanks (UST) are a major source of groundwater contamination. The 1984 RCRA Amendments gave the EPA responsibility to develop a program to minimize this problem and penalties to enforce it. The various states, in turn, have established requirements such as corrosion control and overflow prevention for existing tanks. These matters are discussed in more detail later in this book in Chapter 11.

8.2 GROUNDWATER CONSIDERATIONS IN NEPA STUDIES

NEPA type projects should meet certain requirements to ensure that they do not result in groundwater contamination. These requirements will necessitate answers to the following typical questions:

- Does the project include injection of contaminants into the ground?
- Will products that could contaminate groundwater be stored underground or used in some other manner that could contaminate groundwater?

- To what extent will totally impervious surfaces be constructed during the project?
- Will run-off from impervious surfaces be drained into on-site holding systems?
- Are construction activities likely to contaminate groundwater or to disturb clay layers that seal off existing contaminants from the aquifers below them?
- Is the groundwater a present or future source of drinking water? What is its water quality?

Data sources typically used to develop baseline groundwater condition descriptions include:

- USGS groundwater papers.
- State geological survey.
- State water resource agencies.
- State and local public health departments.
- USGS hydrologic atlases.
- EPA reports.
- Permittee/grantee groundwater investigations.

When more localized or detailed data are required, one may utilize available wells or subcontract the drilling of test wells.

The quantity of groundwater associated with an aquifer can be determined by techniques such as soil conductance measurements and resistivity surveys. Pump tests and slug tests are performed to determine aquifer characteristics such as transmissibility and storage coefficients. For shallow water tables, field methods include monitoring wells and piezometer installation for direct observation and measuring rates of groundwater level increase or dissipation of added water. The NEPA analysis also estimates the reduction of aquifer recharge areas as a result of development and the resultant effects on groundwater quantity.

In determining the quality of groundwater, one utilizes records from well sampling published by USGS and federal, state, and local public health authorities. Supplementary investigations may be conducted by sampling wells, springs, and recharge areas through the installation of peizometers and water quality monitoring instruments in observation or collection wells. Physical water quality criteria are measured at the sampling point (temperature of sample) or evaluated in the laboratory (turbidity, color, odor, and taste).

Various methods may be used to determine groundwater impacts according to specific needs. The appropriate methodology depends in part on the significance of groundwater impacts as determined during scoping of the project, the geology of the study area, the present and future demands on the groundwater resource, and the primary and/or secondary impacts of the proposed action or alternatives on the groundwater resource. In many NPDES projects, the volume of groundwater consumption has been a major issue, as has been saltwater intrusion along coastal areas. In these instances, it is necessary to calculate cumulative groundwater consumption to determine impacts on the aquifers and the effects on other users.

The quality of groundwater at or near sites proposed for development may be sensitive to additional chemical or hydrogeologic stresses imposed by the activity. Groundwater quality monitoring programs in conjunction with quantitative groundwater studies therefore may be required.

Other uses of groundwater which fall between the two extremes of human consumption and deep-well injection are dependent on a variety of water quality criteria. Each NEPA study must address the effects of the proposed project and alternatives on regional groundwater supplies.

Significant aquifer recharge areas are identified based on surface drainage patterns, soil types, subsurface strata, aquifer permeabilities, and groundwater flows. Much of the information collected on topography, soils, and other geologic characteristics of a study area is utilized in this analysis. Estimates of sustainable yield from an aquifer are derived from an analysis of the regional water budget. The precise locations of significant recharge areas and important aquifers are graphically delineated on maps of the study area for the EIS.

To address the effects on groundwater quality of a proposed project, one must first analyze the effects of weathering on any exposed material on the site to determine the nature of the leachate that is produced. The chemical characteristics of the leachate are identified and its effect on existing groundwater (both near-surface and deep aquifers) is assessed. It is also necessary to assess the effects on groundwater quality of dewatering operations and the potential for degradation of surface water quality by groundwater discharges (Conservation Foundation, 1987).

Mitigating measures are primarily preventive in nature. The most effective measures are those that prevent groundwater contamination by not permitting the operation that would contaminate the groundwater to proceed. The use of proper liners for ponds or basins containing possible contaminants is one practical mitigating measure, while double liners of plastic or the use of impermeable clay liners for landfill operations also may be effective. Cleanup of contaminated groundwater is a lengthy and expensive process which may or may not be effective, depending upon the particular situation.

REFERENCES

Briggs, G. F., Developing groundwater resources, in *Handbook of Water Resources and Pollution Control,* Gehm, H. S. and Bregman, J. I., Eds., Van Nostrand Reinhold, 1976, 300.

Groundwater protection, Conservation Foundation, Washington, D.C., 1987.

Salley, W. B., Preliminary estimates of water use in the United States, 1995, U.S. Geological Survey Open-File Report 97-654, Reston, VA, 1997.

9 The Man-Made Environment: Air

This chapter starts with a discussion of climatology, which is necessary as background input into the determination of the impacts of a specific project on the air quality of a region. The chapter then goes on to discuss air pollution per se.

9.1 CLIMATOLOGY

Climate may have a direct and important bearing upon a project that is the focus of an EIS. Depending upon the activity that an EIS is addressing, if alternative locations for a project are geographically widely separated, climate may be a determining factor in the selection of a preferred location site. Climate, including temperature and humidity, controls the growing of crops, the range of plants and animals, the emergence of insects, and the settling of people. Many persons have strong opinions related to abundant rainfall, or the potential for more than an occasional tornado or hurricane. Such opinions, if held by many in the potential project's local area, could have an influence on the success of an EIS project during its early operation stage.

In developing the narrative discussion on climate for an EIS, there are two questions that begin the thought process:

1. What potential effects will prevailing climate have upon the operation of the proposed project or upon the persons who are responsible for the project's success?
2. Is there a potential that waste emissions from the project may influence the prevailing climate and produce secondary concerns such as fog on nearby transportation routes, or any other direct or indirect concerns?

If the answer to either of the preceding questions is yes or maybe, the potential effects should be detailed and discussed.

The annual summary of local climatological data for a particular location, as developed by the National Oceanic and Atmospheric Administration (NOAA), provides plots of daily temperature, precipitation, and sunshine for the year of preparation. It also provides numeric monthly data on temperature and extremes, degree days, percent possible sunshine, sky cover, precipitation, snow and ice, thunderstorms, fog, relative humidity, and wind information for the year of preparation. For similar information, another table provides the normals, means, and extremes, with the extremes associated with the historic year of their occurrence. A 30 year historic record is provided for precipitation, average temperature, heating degree days, cooling degree days, and snowfall. The summary concludes with a narrative discussion of climate for the location addressed. Provided that the location discussed in the weather station information is in reasonable proximity to the EIS project location or preferred location, this useful information could provide the basis for summarizing climatic conditions for an EIS.

Daily climatic data that may be important for the EIS discussion may not be on hard copy, but are available at weather stations, especially those located at airports.

Climatic conditions are integral to studies of air quality effects. For that reason, NEPA studies usually contain a subsection in the section on the existing status of the affected or natural environment that discusses climatology. When examining the impacts of the project in a subsequent part of the NEPA document, climatology is a part of the section dealing with the project effects on air quality. In order to understand air quality effects, it thus is first necessary to have a detailed knowledge of the climatology of the region in which the air quality impacts are going to occur.

A comprehensive listing of climatology should include a discussion of the following factors for air:

1. Ambient conditions:
    a. Temperature—mean monthly values, high and low for year, and daily temperature range.
    b. Precipitation—amount and distribution on a monthly basis, differentiate between rain and snow, present annual high and low records for rain and snow, and mean annual values for rain and snow.
    c. Relative humidity on a monthly basis.
    d. Winds—speed, direction, and so on, on a monthly basis.
2. Storms:
    a. Information frequency, intensity, direction, and so on.
    b. Fogs—these obviously will affect air pollution. Fogs can have particularly dramatic effects.
3. Inversions:
    a. The frequency of inversions in the region and locale, past effects, and so on, and dispersion characteristics. These factors are particularly important in large metropolitan areas such as Los Angeles and Denver.

Climatological data usually are fairly readily available. On a national basis, they may be obtained from the NOAA, National Climatic Center, Asheville, NC. More detailed local information may be obtained form local sources and particularly from local airports. Information also frequently is available from local and state air pollution control agencies as well as regional planning agencies.

In the section on the possible impacts of a project on air quality, it will be seen that the most common techniques involve the use of mathematical models of pollutant dispersion. These models use meteorological data as inputs and yield estimates of pollution concentrations at various locations and heights at outputs. Correct meterological data, especially with regard to wind directions and speeds, obviously will play an important role in determining whether or not a proposed project will violate existing area or regional air quality standards.

9.2   AIR POLLUTION

The material that follows is divided into two topics:

- The legal requirements that must be described and with which the degree of compliance of a specific project must be shown in the EIS.
- The methodology utilized to predict impacts of projects.

Before discussing each of these topics, it should be noted that the Clean Air Act and the effort it has engendered to clean up this nation's air is working. According to the Environmental Protection Agency (1996), pollution concentrations for all of the six major air pollutants have declined as follows between 1987 and 1996:

- *Ozone* decreased 15 percent.
- *Lead* in the air decreased 75 percent.
- *Sulfur dioxide* decreased 37 percent.
- *Carbon monoxide* decreased 37 percent.
- *Nitrogen dioxide* decreased 10 percent.
- *Particulates* (dirt, dust, and soot) decreased 25 percent.

9.3   LEGAL REQUIREMENTS

Legal requirements that affect the preparation of the air quality sections of EIS are based almost entirely on the Clean Air Act, as well as analogous state requirements deriving directly from it in most cases. The discussion that follows will proceed accordingly.

The Clean Air Act was passed in 1970, amended in 1977, and amended again in 1990. The Act is designed to protect and enhance the nation's air quality, as well as to safeguard public health and welfare and the productive capacity of its people. The Act is divided into three titles:

- Title I deals with control of pollution from stationary sources.
- Title II deals with control of pollution from mobile sources.
- Title III addresses general administrative matters.

The Act requires the EPA to:

1. Promulgate national ambient air quality standards (NAAQS) for certain pollutants to protect the public health (primary NAAQS) and to protect the public welfare (secondary NAAQS).
2. Establish procedures for collecting and interpreting air quality data.
3. Develop emission standards and control technology guidelines relating to the control of emissions from stationary sources of air pollutants (such as factories, power plants, refineries, and other industrial facilities).
4. Develop emission and fuel standards for motor vehicles.

The EPA also supervises state air pollution control efforts.

9.4   NAAQS

The Clean Air Act of 1970, as amended in 1977, required that the EPA establish primary and secondary air quality standards for each of the six common air pollutants (criteria pollutants): carbon monoxide, lead, nitrogen dioxide, ozone, particulates, and sulfur dioxide. For each of the air quality standards, the EPA was to:

1. Set a maximum concentration level.
2. Specify an averaging time over which the concentration is to be measured.
3. Identify how frequently the time-averaged concentration may be violated per year.

For the ozone standard, for example, the concentration level has been set at 0.08 parts of ozone (O_3) per million parts of air (or 0.08 ppm), daily maximum 8 hour (h) average, not to be exceeded at each air quality monitor on a three-year average of the fourth highest daily maximum 8-h O_3 concentration.

In the most recent regulations concerning nitrogen oxide (NO_x) emissions, the EPA has decided that 22 states must cut their NO_x emissions by 1.6 million tons a year by 2005. The EPA requires them to submit plans for emission reductions by 1998, have controls in place by 2002, and achieve the goals by 2005.

The EPA says that most reductions can come from power plants. The plan requires states to cut their NO_x emissions by 35 percent of what they would otherwise be in 2007, or under 2.9 million tons. The EPA released guidance in 1998 to establish NO_x emissions trading programs for utilities. The EPA said that states may be able to generate a pool of reductions they could use to avoid certain requirements for the construction of new sources, specifically the new ozone standards.

The primary standards are:

1. Uniform across the country, though the states may impose stricter standards if they wish.
2. Set with an adequate margin of safety for those especially vulnerable to pollution, such as the elderly and children.
3. Set without regard to the costs or technical feasibility of attainment.

A deadline of 1972 was initially set for achieving the primary air quality standards. It was later extended for ozone and carbon monoxide, first to 1975, then to 1982, and to 1987.

The secondary standards are intended to prevent damage to soils, crops, vegetation, water, weather, visibility, and property. No deadlines have been set for attaining the secondary standards, but the Act calls for their attainment as expeditiously as practicable.

Each state is required to adopt a plan, called a state implementation plan (SIP), that limits emissions from air pollution sources to the degree necessary to achieve and maintain the NAAQS. The SIP provides emission limitations, schedules, and timetables for compliance by stationary sources. The Act focuses on major stationary sources or major modifications of existing sources. Major sources are defined as sources which emit, or have the potential to emit, more than a prescribed amount of a designated pollutant.

States are also required to adopt measures to prevent significant deterioration of air quality (PSD) in clean air areas. When an SIP is approved by the EPA, it is enforceable by both the federal and state governments.

9.4.1 AIR QUALITY DATA COLLECTION AND INTERPRETATION

The EPA established the procedures for collecting air quality data. Each of the nation's 242 air quality control regions—geographic areas that share common air quality concerns—places one or more air quality monitors at various sites using these procedures. The monitors record hourly concentration-level readings. The EPA then uses the data to define each region as an attainment (clean) or nonattainment (polluted) area for each pollutant. A region can be a nonattainment area for one pollutant and an attainment area for others.

To determine whether an area is complying with the contaminant standards, the EPA counts the number of times the area exceeds the limits. This occurs when the level of the contaminant is above the standard level for 1 h or longer at 1 or more monitors during a 24 h day. The standard allows a certain number of times the area may exceed the limit at each monitor on separate days over any three-year period. As soon as any single monitor registers more than the allowable number of times, the area is classified as being a nonattainment area (Clean Air, 1990).

9.4.2 REGULATION OF STATIONARY POLLUTION SOURCES

The Clean Air Act establishes two major regulatory programs for stationary sources. In the first, the new source performance standards (NSPS) program establishes stringent emissions limitations for new sources in designated industrial categories regardless of the state in which the source is located or the air quality associated with the area. These new stationary source standards directly limit emission of air pollutants (or in the case of the pollutant ozone, its precursors, that is, the chemicals that react to form ozone). The standards apply to categories of sources. For example, the EPA has set emission limits for new petroleum refineries.

The second program, the national emissions standards for hazardous air pollutants (NESHAP), regulates emissions of toxic pollutants for which no NAAQS is applicable, but which cause increases in mortality or serious illnesses (U.S. EPA, 1989).

For existing sources, Section 109 of the Act requires that the EPA adopt national ambient air quality standards for so-called criteria pollutants to protect public health and welfare. There are six criteria pollutants: particulate matter, sulfur dioxide, carbon monoxide, ozone, nitrogen dioxide, and lead.

9.4.3   CONTROL TECHNOLOGY GUIDELINES

The EPA has designated all areas of the country as either attainment or nonattainment for each of the criteria pollutants. SIPs must assure attainment of NAAQS by prescribed dates. SIPs must meet federal requirements, but each state may choose its own mix of emission controls for sources to meet the NAAQS.

The EPA issues control technique guidelines to help states choose the right controls for existing stationary sources in nonattainment areas. The guidelines suggest control technology that will meet the Clean Air Act requirement for the use of reasonably available control technology (RACT) for these existing sources. RACT is defined as the most stringent controls achievable, considering cost and available technology.

In addition, the Clean Air Act calls for:

1. Installation in attainment areas of the best available control technologies (BACT), defined as the maximum degree of emission control achievable, considering, energy, environmental, and economic impacts.
2. Installation in nonattainment areas of the lowest achievable emissions rate (LAER). (Clean Air, 1990). That is, new plants must install equipment that limits pollution to the lowest rate of any similar factory anywhere in the country.

For new or modified stationary sources of air pollution, the Act requires the EPA to promulgate uniform federal new source performance standards (NSPS) for specific pollutants in industrial categories based upon adequately demonstrated control technology. Rather than tying control levels to National Ambient Air Quality Standards, Congress required the EPA to base these uniform emission standards on strictly technological considerations.

The owner or operator of a new or modified source must demonstrate compliance with an applicable new source performance standard within 180 days of the initial start-up of the facility and at other times as required by the EPA. The EPA has primary authority for enforcement of NSPS unless authority is delegated to the states. In such cases, the EPA and the states have concurrent enforcement authority.

For new sources or modification of existing sources, the Clean Air Act requires a preconstruction review. One of the EPA's requirements for this review is that in nonattainment areas, pollution from existing stationary sources be reduced enough to more than compensate for the additional pollution expected from the new source. At present, the EPA requires an offset of roughly 120 percent. This means that a company wanting to build has to purchase emission offsets, that is, pay for emission reductions in someone else's plant if it cannot offset the increase at one of its own plants. In another variation of this approach, the EPA has devised an emissions trading policy—called the bubble policy. One example of its application relates to plants that want to modify their facilities. A plant can do that (and avoid a new source review) by showing that total emissions under an imaginary bubble covering itself or a group of plants will not exceed a predetermined amount despite the modification. A plant may be able to achieve this by altering the emission controls on existing parts of its operations.

The bubble policy may also apply to existing plants faced with meeting new emissions reduction requirements. Within the bubble, these plants may make adjustments so long as the new emissions goal is met (Clean Air, 1990).

9.4.4   PREVENTION OF SIGNIFICANT DETERIORATION (PSD) OF AIR QUALITY

Part C of Title I of the Act, prevention of significant deterioration (PSD) of air quality, applies in all areas that are attaining the national ambient air quality standards where a major source or modification is proposed to be constructed. The purpose of Part C is to prevent the air quality in relatively clean areas from becoming significantly dirtier. A clean air area is one where the air quality is attaining the ambient primary and secondary standard. Designation is pollutant-specific so that an area can be nonattainment for one pollutant, but clean for another. It establishes three classifications or geographical areas for proposed emitters of sulfur dioxide and particulate matter:

- Only minor air quality degradation allowed—Class I.
- Moderate degradation allowed—Class II.
- Substantial degradation allowed—Class III.

In no case does PSD allow air quality to deteriorate below secondary air quality standards. Baseline is the existing air quality for the area at the time for which the first PSD is applied. Increments are the maximum amount of deterioration that can occur in an attainment area over the baseline. Increments in Class I areas are smaller than for Class II, and Class II increments are smaller than for Class III areas.

For purposes of PSD, a major emitting source is one of 26 designated categories which emits or has the potential to emit 100 tons per year of the designated air pollutant. A source that is not within the 26 designated categories is a major source if it emits more than 250 tons per year.

Any proposed major new source or major modification is subject to preconstruction review by the EPA, by a state to whom the program is delegated or by a state which has adopted PSD requirements in its SIP, so that a permit for increases will not be exceeded. The permit describes the level of control to be applied and what portion of the increment may be made available to that source by the state. Where the EPA has delegated such review, the EPA and the state have concurrent enforcement authority (U.S. EPA, 1989).

Nonattainment areas are those which are not in compliance with national air quality standards. For a proposed source which will emit a criteria pollutant in an area where the standards are presently being exceeded for the pollutant, even more stringent preconstruction review requirements apply. This review is the primary responsibility of the state where the source is proposed to be constructed, with overview authority vested in the EPA.

New construction of major sources or major modifications in a nonattainment area (NAA) is prohibited unless the SIP provides for the following:

- The new source will need an emission limitation for the nonattainment pollutant that reflects the lowest achievable emission rate.
- All other sources within the state owned by the subject company are in compliance.
- The proposed emissions of the nonattainment pollutant are more than offset by enforceable reductions of emissions from existing sources in the nonattainment area.
- The emission offsets will provide a positive net air quality benefit in the affected areas.

The applying source in the NAA must therefore obtain a greater than 1 : 1 reduction of the pollutant or pollutants for which the area has been designated nonattainment. Emission offsets from existing sources may need to be obtained, especially if the new source will have emissions that would exceed the allowance for the NAA. In these situations, the source would need to obtain enforceable agreements from other sources in the NAA, or from its own plants in the NAA.

9.4.5 NATIONAL EMISSION STANDARDS FOR HAZARDOUS AIR POLLUTANTS (NESHAP)

Section 111 of the Clean Air Act defines hazardous air pollutants as those for which no air quality standard is applicable, but which are judged to increase mortality or serious irreversible or incapacitating illness. National emissions standards for hazardous air pollutants (NESHAP) standards are based on health effects with strong reliance on technological capabilities. These standards apply to both existing and new stationary sources. The NESHAP program can be delegated to any qualifying state.

Under NESHAP, no person may construct any new source unless the EPA determines that the source will not cause violations of the standard. For existing sources, a standard may be waived for up to two years if there is a finding that time is necessary for installation of controls and that steps will be taken to prevent endangerment of human health in the interim (U.S. EPA, 1989).

9.4.6 REGULATION OF MOTOR VEHICLES AND FUELS

While the EPA is responsible for regulating the manufacture of new motor vehicles nationwide, states may control motor vehicle emissions by methods that do not require vehicle modification. California is the only state that may require pollution control equipment on motor vehicles. Initially, the Clean Air Act made the EPA responsible for setting motor vehicle emission standards to accomplish emission reductions prescribed in the Act. Those reductions have been achieved. The EPA also has the authority to regulate fuels; for example, it has proposed a rule to lower gasoline volatility.

SIPs detail state strategies for emission reductions to meet NAAQS. The EPA must approve the state's pollution-reduction plans. If the EPA finds a state's plan inadequate, the EPA can revise it or require that the state do so, or the EPA can impose sanctions—such as construction bans and federal funding restrictions. The EPA can impose sanctions if it finds that a state is not implementing its approved plan. Furthermore, if a plan is inadequate, the EPA can prescribe a federal implementation plan (FIP) for the state (Clean Air, 1990).

9.5   THE 1990 CLEAN AIR ACT AMENDMENTS

In late 1990, Congress passed the Clean Air Amendments that were designed to focus on acid rain, urban air pollution, and toxic air emissions. The amendments also added programs to address accidental releases of toxic air pollutants.

Key provisions and milestones of the amendments are as follows (U.S. EPA, 1990): Titles I, III, IV and V of the 1990 Amendments relate to stationary sources. Title I addresses nonattainment areas. For the pollutant ozone, the new law establishes nonattainment area classifications for metropolitan areas ranked according to the severity of the air pollution problem. These five classifications are marginal, moderate, serious, severe, and extreme. The EPA assigns each nonattainment area one of these categories, thus triggering various requirements that the area must comply with in order to meet the ozone standard. Marginal areas, for example, are the closest to meeting the standard. They will be required to conduct an inventory of their ozone-causing emissions and institute a permit program. Moderate areas and above must achieve 15 percent volatile compounds reduction within 6 years of enactment. For serious and above, an average of 3 percent volatile organic compounds reduction per year is required until attainment. For the city of Los Angeles, for example, this translated to a 20 year ozone reduction program to achieve attainment (U.S. EPA, 1990). The law establishes similar programs for areas that do not meet the federal health standards for the pollutants carbon monoxide and particulate matter.

A summary of key time tables for Title I requirements follows:

- States bring their existing control requirements into line with federal standards within six months.
- Designation and classification of areas completed with 480 days.
- EPA guidance for state implementation plans (SIPs) issued within one year.
- Complete SIPs submitted by states within three years.
- First group of guidance documents for retrofit controls on existing sources issued within two years.

Title III address emissions of toxic pollutants. The amendments list 189 hazardous air pollutants. Within one year, the EPA must list the source categories that emit one or more of the 189 pollutants. Within two years, the EPA must publish a schedule for regulation of the listed source categories. For all listed major point sources, the EPA must promulgate maximum achievable control technology standards. These standards must address 40 source categories plus coke ovens within 2 years, 25 percent of the remainder of the list within 4 years, an additional 25 percent in 7 years, and the final 50 percent in 10 years. The maximum achievable control technology regulations are emissions standards based on the best demonstrated control technology and practices in the regulated industry. For existing sources, they must be as stringent as the average control efficiency or the best controlled 12 percent of similar sources. For new sources, they must be stringent as the best controlled similar source. As a consequence, the EPA has drafted a list of 40 air toxics that will be the basis for air toxics standards, vehicle fuel standards, and state air pollution control requirements that target large urban areas. This effort is known as the urban air toxics strategy. Available toxicity, ambient air monitoring and emissions inventory data, and results from exposure and risk assessment studies were used to develop the list.

At the time of the preparation of this book, the EPA had indicated that it would publish a notice announcing the availability of the preliminary data. Meanwhile, copies of the inventory report can be downloaded from www.epa.gov/ttn/uatw/-112kfac-html.

Title IV is designed to reduce acid rain. It is intended to result in a permanent 10 million ton reduction in sulfur dioxide emissions per year from 1980 levels. The first phase, effective January 1, 1995, was to affect 110 powerplants and provide them with certain reduction allocations. The second phase will become effective January 1, 2000 and will affect 2000 utilities. In both phases, affected sources will be required to install systems that continuously monitor emissions in order to track progress and assure compliance. The law allows utilities to trade emission allowances with their system or buy or sell allowances to and from other affected sources. A summary of key time tables for Title IV follows:

- Allowances for the first phase of the control program are issued within 12 months; second phase is issued six months later.
- Allowance trading regulations issued within 18 months.
- Continuous emissions monitoring regulations issued within 18 months.

Title V establishes a clean air permit program similar to the NPDES permit program in water. The EPA must issue program regulations within one year. Within three years, each state must submit to the EPA a permit program meeting these regulatory requirements. After receiving the state submittal, the EPA has one year to accept or reject the program. The EPA must levy sanctions against a state that does not submit or enforce a permit program.

All sources subject to the permit program must submit a complete permit application within 12 months. The state permitting authority must determine whether or not to approve an application within 18 months of the date it receives the application. Each permit issued to a facility will be for a fixed term of up to five years. The new law establishes a permit fee system whereby the state collects a fee from the permitted facility to cover reasonable direct and indirect costs of the permitting program.

Title II, which relates to mobile sources of air pollution, has the following requirements for new programs standards and regulations (U.S. EPA, 1990):

- Within 1 year, new inspection and maintenance programs must be established in over 50 cities.
- Within 2 years, enhanced inspection and maintenance programs must be established in over 70 cities.

- New tailpipe regulations must be issued within six months; standards were to begin phasing in with the 1994 model year.
- Reformulated gasoline regulations were to be issued by the EPA within one year.

Finally, Title VI relates to stratospheric ozone and global climate protection. The law requires a complete phaseout of certain chemicals that affect the ozone layer. Leading up to a phaseout, there will be stringent interim reductions placed upon the specific chemicals. Within 60 days of enactment, the EPA must list all regulated substances along with their ozone-depletion potential, atmospheric lifetimes, and global warming potential.

9.6   IMPACT PREDICTION METHODOLOGY

The prediction of primary and secondary air quality impacts, along with a discussion of compliance with the Clean Air Act, represents the essence of the air quality sections of an EIS.

Primary air quality impacts are associated with the general degradation of the ambient air quality and the resultant effects on health, vegetation, wildlife, and general environs. Secondary air quality impacts are caused by the overall effects of induced population growth and associated activities.

9.6.1   BASELINE DATA

The first phase of the EIS effort involves the establishment of baseline conditions, that is, the air quality situation before the project or program is undertaken.

Climate conditions are integral to studies of air quality effects. That is the reason for the inclusion of climatology in this chapter. Existing climatic conditions are described in terms of monthly precipitation, wind data, mean monthly temperatures, daily temperature range, mean snowfall, and heating/cooling degree days. Climatological data are available from the National Oceanic and Atmospheric Administration. They also are obtained from local sources, for example, airports.

As the first phase of the data gathering effort, one lists the federal and state air quality standards for specific pollutants that apply to the sites under consideration. This includes both the primary (public health protection) and secondary (public welfare protection) standards. Wherever duplication exists between federal and state standards, the more restrictive standard applies.

The EIS preparer then obtains and shows all available information on the present concentrations of the following pollutants near the sites:

- Carbon monoxide.
- Nitrogen dioxide.
- Photochemical oxidants.
- Total suspended particulates.
- Sulfur dioxide.

In establishing baseline conditions, one relies primarily on readily available data including that obtainable from the following sources:

- NOAA, National Climatic Center.
- State air pollution control agencies.
- EPA.
- Local air pollution control agencies.
- Local public health agencies.

These numbers are compared with the standards that were presented earlier. Particular attention is paid to the question of nonattainment areas, especially for ozone.

There are a number of data bases available that may be helpful in establishing baseline conditions. These include the following (EPA, 1998):

AIRSweb—provides easy access to key measurements of air pollution that the EPA uses to assess the Nation's air quality, including air quality measurements from 4000 air monitoring sites across the nation for the past five years and air pollutant emissions and regulatory compliance status for 9000 point sources regulated by the EPA.

AIRS-Aerometric Information Retrieval System—a computer-based repository of information about airborne pollution in the United States and various World Health Organization (WHO) member countries.

AIRS Executive Software—personal computer data base that contains a select subset of data extracted from the AIRS data base.

Applicability Determination Index—a data base that contains memoranda issued by the EPA on applicability and compliance issues associated with the new source performance standards (NSPS), national emissions standards for hazardous air pollutants (with categories for both NESHAP, Part 61, and MACT, Part 63), and chlorofluorocarbons (CFC).

9.6.2   IMPACT CALCULATIONS

Based on the baseline conditions survey, the air quality growth constraints in the planning area are described in terms of the acceptable ambient air pollutant increases, if any, for the criteria pollutants. The proposed project then is evaluated in terms of the applicable ambient standard averaging times. The margins for increases would be the lesser values of those allowed by the prevention of significant deterioration (PSD) increments or those allowed by the national ambient air quality standards. For any cases in which little or no margin for increase exists, the potential applicability of emission offset techniques or design optimization analysis would be addressed.

Computer modeling is used for evaluating primary impacts from large projects (emissions of over 50 tons per year, 1000 lb per day, or 100 lb per h of pollutants) and/or from complex sources in sensitive areas. The first step is to compile all necessary data to be used in the computer model and in the ultimate evaluation of the results of the modeling effort. They include ambient air quality, meteorological, operations, and point-source emissions data for other area sources. Next, a computer model or series of models is selected. The computer programs are determined by the needs and characteristics of each individual project.

After the most appropriate computer model(s) have been selected, the meteorological data have been analyzed for appropriate fit, and the operational parameters, and project emission characteristics have been selected, a computer analysis is performed which is made under worst case meteorological and operational conditions specific to the proposed project. The results are analyzed for PSD and NAAQS compliance. If they demonstrate that the proposed project is in compliance with applicable local, state, and federal regulations, the results are summarized in both tabular and narrative form. If the analysis demonstrates that the proposed project would not be in compliance, an optimization analysis should be designed in which variables that are changed easily such as stack height, stack diameter, and exhaust temperature are evaluated. This analysis consists of performing a sensitivity study on the interrelationship between those variables and the resultant downwind concentrations.

Odor and aerosol problems and possible solutions to them vary widely. One must determine the cause of the problem before attempting to evaluate its impacts and possible mitigating measures. For the purposes of this discussion, we will concentrate on odors from municipal and industrial wastewater treatment plants, since they are the most frequent ones encountered.

Odors and aerosols released from municipal and industrial wastewater treatment facilities can constitute a significant problem at the facilities and to nearby residences. Two methodologies frequently are used to evaluate special odor problems: gas chromatography (GC) fingerprinting and the pollution or ammonia rose technique. The GC fingerprinting methodology compares the results of GC analyses from air at the odor source and from an area that apparently is being affected by the odor source. If the two samples have the same GC characteristics, it is concluded that the second site is being impacted by the odor source. To develop an ammonia rose, one analyzes historic or actual site wind directional data and reported ammonia concentrations and frequencies from monitoring sites surrounding the project site. The direction, concentration, and frequency of occurrence of ammonia concentrations would be reported.

Two methodologies are used to evaluate aerosol problems. The first involves computer modeling of aerosol drift to determine projected downwind concentrations. The second methodology to evaluate aerosol problems involves ambient modeling and is used to determine actual downwind concentrations from a project site. This type of study requires in-depth meteorological measurements and elaborate air quality sampling. Continuous meteorological data, including wind direction and velocity, are collected. Upwind samples are taken and used as a control or background sample, whereas downwind samples would be indicative of concentrations produced from the project.

To evaluate construction-related air quality impacts (fugitive dust) one develops a worst case site emissions day. This is accomplished by estimating sources and amounts of dust generated and what kinds and how many construction vehicles would be used on the day(s) of highest construction activity. The number and types of construction vehicles would be compiled in tabular form and emission rates for each vehicle would be determined based on rates in EPA's Publication AP-42. These site emission rates then would be evaluated in relation to the size of the construction site, meteorological data, ambient air data, abatement measures used (such as spraying of water on dust, etc.), and privately owned vehicles used in commuting to and from the site. If it is determined that the potential primary impacts are significant, then an area source computer model could be considered.

The secondary impacts of new housing, increased vehicle and home heating emissions, and the overall effects of population growth on ambient air quality are estimated using appropriate data for domestic fuel consumption (for space and hot water heating), electrical energy usage, and vehicular mileage, in combination with average land use based on emissions factors. EPA publication EPA-3-76-012b is an important aid in conducting this evaluation.

9.7   POSSIBLE MITIGATION MEASURES

Air quality is a technological area where mitigating measures are readily available. Dust raised during construction can be controlled by the following approaches:

- Designing the construction activities in such a manner that a minimum amount of fugitive dust will be created and will be kept within the project boundaries by barriers or absorbent materials.
- Scheduling construction to avoid dry seasons and higher inversion periods.
- Spraying water on the soil and excavated material to keep it on the ground.
- Graveling access roads.
- Covering soil and debris piled in open trucks.

Air contaminants in the exhausts of vehicles used in construction are relatively minor and generally are not considered.

Air contaminants generated by the project itself usually may be mitigated by the use of one of the following methods:

- Equipment that traps the contaminants—precipitators, bag house collectors, limestone slurries, etc.
- High smoke stacks (where allowed by the regulatory agency).
- Use of minimally polluting fuels (e.g., low sulfur).
- Tradeoffs for capacity available in other areas.
- Plant operation during hours when other pollutant generators are not working.

Mitigating measures for secondary impacts will vary with the type of impact. Thus, for example, effects of increased vehicle emissions may be mitigated to some extent by proper highway and traffic control design. In general, the induced population growth results in a variety of emissions that cannot be controlled easily in a direct manner. Land-use planning and zoning becomes an indirect mitigating measure.

Mitigating measures for odors tend not to be very successful. Absorption, venting enclosures, and industrial perfumes all have had some degree of success. It is often better to attempt to determine the cause of the odor and to minimize it, for example, location in the case of sludge heaps.

A high-purity oxygen system may be used for the secondary treatment process because of its greater resistance to upsets and subsequent odor problems. Egg-shaped digesters may be installed because of their self-cleaning properties, which minimize the need for digester cleaning—a potentially odorous operation.

Pathogenic aerosol drift problems for several wastewater treatment facility projects have been resolved by the author of this book utilizing land application. For application type alternatives proposed in Scottsbluff, NE and Bimidji, MN, mitigating measures used to reduce aerosol drift included the following:

- The use of buffer zones between spray irrigating areas and receptor locations.
- The planting of dense screening vegetation.
- The restriction of spray irrigation operations to daylight hours.

Aerosol drift associated with lagoon wastewater treatment facilities normally is best controlled by determining optimum placements for the facilities.

REFERENCES

Clean air working progress: today and for the future, National Association of Manufacturors, Clean Air Working Group, Washington, D.C., 1990.

Basic inspectors training course: fundamentals of environmental compliance inspection, U.S. Environmental Protection Agency, Office of Compliance Monitoring, Washington, D.C., 1989.

The Clean Air Act Amendments of 1990, Summary materials and fact sheets, U.S. Environmental Protection Agency, 1990.

Growth effects of major land use projects, in *Complication of Land Use Based Emission Factors,* Vol. II, U.S. Environmental Protection Agency, Publication No. US EPA-3-76-012b.

Media specific tools, U.S. Environmental Protection Agency, Databases and Software–Media Information, Washington, D.C., 1998.

National air quality and emissions trends report, U.S. Environmental Protection Agency, Office of Air Quality Planning and Standards, 1996.

U.S. Environmental Protection Agency, Publication No. AP-42.

10 The Man-Made Environment: Noise

10.1 IMPACTS

Noise at levels that may be objectionable in terms of health or nuisance effects as considered in an environmental impact study generally will occur as a result of one of the following activities:

- Construction and plant operation.
- Vehicular traffic.
- Aircraft.
- Population growth and urbanization.

The concern about noise is directly related to its negative impacts upon humans and animals. Newman and Beattie (1985) have summarized these effects as follows:

- Annoyance.
- Permanent or temporary hearing loss.
- Speech interference.
- Sleep interference.
- Health impacts.
    - Cardiovascular effects.
    - Achievement scores.
    - Birth weight.
    - Mortality rates.
    - Psychiatric admissions.
- Harm to animals.
- Effects on productivity of domestic animals.
- Vibration of walls and windows.
- Radiation of secondary noise.
- Human physiological response to intense low frequency sound.
- Sonic booms.

While these possible effects were tabulated in regard to aviation noise, they also cover possible effects of noise from other sources such as those mentioned at the beginning of this chapter. Each of these possible effects will be discussed briefly.

Annoyance may be on an individual or community-wide basis. It is probably the most common reaction to unwanted, unpleasant noise. On an individual basis, it provokes an emotional reaction to the noise that may include a physiological response, a concern about the effect of the noise on the individual's health, an anger about the inability to carry on a conversation at normal levels, possible fears associated with the source of the noise, and above all, anger at the intrusion of one's privacy, especially when the noise is heard indoors. Many of those same factors apply to a community reaction to noise. Here the major annoyances may include anger at the destruction of the general ambience which sets the tone for the community, as well as interference with home values, community events (schools, churches, etc.), and outdoor activities.

Hearing loss may be either permanent or temporary. Continuous exposure to high levels of noise will damage human hearing. The upper limit of hearing is about 120 dB (Newman and Beattie, 1985), at which discomfort begins to occur. Pain usually starts at 140 dB with auditory fatigue or acoustical injury eventually being reached. However, even sounds below the 90 to 100 dB range may bring about short-term changes in hearing.

A temporary reduction in hearing acuity is a common effect of noise in industrial or entertainment situations. After exposure to high noise levels for a short time, or moderate noise levels over a long time, some hearing ability may be lost. Recovery of hearing usually occurs within several hours (Newman and Beattie, 1985). Longer exposure to high noise levels may cause a degree of permanent hearing loss or, at a minimum, ringing in the ears.

Speech interference could have been listed as a subcategory under annoyance. The distance at which speech is intelligible can be shortened dramatically by loud noise. *Sleep interference* is properly classified as a health problem. The importance of this effect is shown by the fact that most state and local noise laws have much lower maximum values at night than in the daytime. Other health effects listed in the second paragraph of this chapter are self-evident and do not require further elaboration.

Harm to animals is difficult to quantify since laboratory studies are often quite dissimilar to the real situation. Nevertheless, certain effects are obvious. One may divide the effects into two categories—wild and domestic animals. Wild animals are considered to be those that live in wildlife refuges, national parks, and wilderness areas. In the case of short-time noises, for example, construction, the animals may simply vacate the area. Whether or not they come back again depends upon the nature of the project. However, for continuing noise such as from traffic or aviation, the response of animals appears to be species-dependent and varies from almost no reaction to no tolerance of the sound (Newman and Beattie, 1985).

Some birds will be driven away permanently from nesting areas as a result of a project that brings a human population into the area (e.g., eagles) whereas others do not seem to be affected at all. The same applies to vehicular traffic and aviation flights as well. Newman and Beattie (1985) have described a study by Edwards et al. (1979) which observed 11 different avian species in a test employing helicopters and other aircraft. Exhibit 5 presents the results of the Edwards (1979) study and shows how the various species of birds reacted as the decibel level was increased from 70 to 95 dB.

According to Newman and Beattie (1985), fish appear to have little response to outside noises, even including sonic booms. This apparently is because most of the sound is reflected off the water.

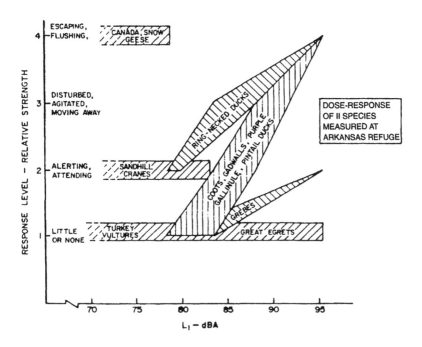

Exhibit 5   Dose-response of 11 species measured at Arkansas refuge. (From Edwards et al., U.S. Department of Transportation, Federal Aviation Administration, 1979. With permission.)

Studies on pigs (Bond et al., 1980) have shown that jet and propeller aircraft sounds from 120 to 135 dB for 12 h, from weaning time to slaughter, showed no effects on feeding or weight gain. A similar study on dairy cattle showed no differences in milk production or any other abnormal effects (Parker and Bayley, 1960). Poultry also have shown no appreciable reaction to aircraft flyover noise and neither have mink (Newman and Beattie, 1985). On the other hand, laboratory studies of rats, mice, monkeys, and rabbits have shown hearing damage when these animals were exposed to high level noise (Newman and Beattie, 1985).

10.2   NOISE LEVEL EXPLANATION

Noise is measured in decibels. This number is equivalent to the sound pressure level. The human ear perceives sound, which is mechanical energy, as a pressure on the ear. The sound pressure level is the logarithmic ratio of that sound pressure to a reference pressure and is expressed as the decibel.

Sound is measured by a meter which reads units called decibels (dB). For highway traffic and other noises, an adjustment, or weighing, of the high- and low-pitched sounds is made to approximate the way that an average person hears sounds. The adjusted sounds are called A-weighted levels (dBA).

The A-weighted decibel scale begins at zero. This represents the faintest sound that can be heard by humans with very good hearing. The loudness of sounds (that is, how loud they seem to humans) varies from person to person, so there is no precise definition of loudness. However, based on many tests of large numbers of people, a sound level of 70 is twice as loud to the listener as a level of 60. This principle is illustrated in Exhibit 6, which was produced by the Federal Highway Administration (FHWA, 1980).

Newman and Beattie (1985) have presented an excellent table that shows typical decibel values encountered in daily life and in industry. That table is reproduced as Exhibit 7. Exhibit 8, which is taken from the EPA (EPA, 1974), shows typical noise levels at construction sites. Noise levels during construction of a facility are temporary but generally are high enough so that precautions must be taken.

The Federal Highway Administration (FHWA, 1973) has developed a set of noise level/land-use relationships that are useful in determining the significance of noise levels in relation to land uses. The values shown in Exhibit 9 are those that generally are found in the indicated land uses and to which highways and other noise-generating activities should be designed.

Decibel levels are not additive, that is, one cannot add a 70 dB noise to a 70 dB setting and obtain 140 dB. Instead, the final result would be between 73 and 74 dB. In addition, noise is three-dimensional in nature because of its sound wave characteristics. Consequently, in projecting noise effects on a specific setting, such as from a highway onto different levels of a nearby house, one must analyze a three-dimensional model and a time of day factor as well. These models are particularly and frequently used for highway traffic and aviation noise effects on buildings.

Highway traffic noise is the largest single source of noise that is considered most frequently in an EIS. This noise is not constant. The noise level changes with the number, type, and speed of the vehicles which produce it. Traffic noise variations can be plotted as a function of time. However, it is usually inconvenient and cumbersome to use such a graph to represent traffic noise. A more practical method is to convert the noise data to a single representative number.

Statistical descriptors are almost always used as a single number to describe varying traffic noise levels. The two most common statistical descriptors used for traffic noise are L_{10} and L_{eq}. L_{10} is the sound level that is exceeded 10 percent of the time. L_{eq} is the constant, average sound level, which, over a period of time, contains the same amount of sound energy as the varying levels of the traffic noise. L_{eq} for typical traffic conditions is usually about 3 dBA less than the L_{10} for the same conditions.

The Federal Highway Administration (FHWA, 1980) has established noise impact criteria for different land uses close to highways. Some of the exterior criteria are illustrated below.

Land use	L_{10}	L_{eq}
Residential	70 dBA	67 dBA
Commercial	75 dBA	72 dBA

If a project causes a significant increase in the future noise level over the existing noise level, it also is considered to have an impact.

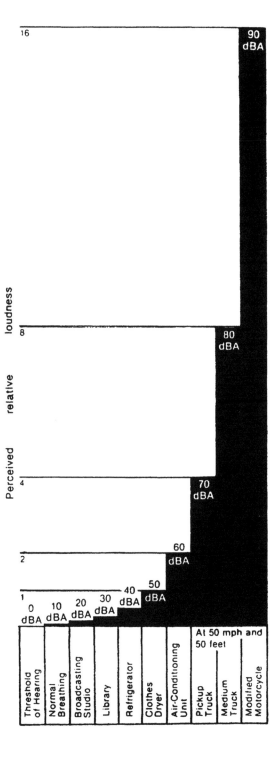

Exhibit 6 Perception of sound levels. (From Federal Highway Administration, 1980. With permission.)

Exhibit 7
Comparative Noise Levels

Typical Decibel (dBA) Values Encountered in Daily Life and Industry	dBA
Rustling leaves	20
Room in a quiet dwelling at midnight	32
Soft whispers at 5 ft	34
Men's clothing department of a large store	53
Window air conditioner	55
Conversational speech	60
Household department of large store	62
Busy restaurant	65
Typing pool (9 typewriters in use)	65
Vacuum cleaner in private residence (at 10 ft)	69
Ringing alarm clock (at 2 ft)	80
Loudly reproduced orchestral music in large room	82

Beginning of Hearing Damage if Prolonged Exposure over 85 dBA	
Printing press plant	86
Heavy city traffic	92
Heavy diesel-propelled vehicle (about 25 ft away)	92
Air grinder	95
Cut-off saw	97
Home lawn mower	98
Turbine condenser	98
150 ft^3 air compressor	100
Banging of steel plate	104
Air hammer	107
Jet airliner (500 ft overhead)	115

Source: From Newman, J. S. And Beattie, K. R., Aviation noise effects, U.S. Department of Transportation, Report No. FAA-EE-85-2, 1985. With permission.

Exhibit 8
Typical Ranges of Energy-Equivalent Noise Levels in dBA at Construction Sites

Phase	Domestic Housing		Office Building, Hotel, Hospital, School, Public Works		Industrial Parking Garage, Religious, Amusement and Recreation, Store Service Station		Public Works Roads and Highways, Sewers, and Trenches	
	I	II	I	II	I	II	I	II
Ground clearing	83	83	84	84	84	83	84	84
Excavation	88	75	89	79	89	71	88	78
Foundations	81	81	78	78	77	77	88	88
Erection	81	65	87	75	84	72	79	78
Finishing	88	72	89	75	89	74	84	84

Note: I represents all pertinent equipment present at site and II represents the minimum required equipment present at site.

Source: Information on levels of environmental noise requisite to protect public health and welfare with an adequate margin of safety, U.S. Environmental Protection Agency, Publication 550/9-74-004, Office of Noise Abatement and Control, Washington, D.C., 1974. With permission.

Exhibit 9
Noise Level/Land-Use Relationships

Land-Use Category	Design Noise Level-L_{10}	Description of Land-Use Category
A	60 dBA (exterior)	Tracts of land in which serenity and quiet are of extraordinary significance and serve an important public need, and where the preservation of those qualities is essential if the area is to continue to serve its intended purpose. Such areas could include amphitheaters, particular parks or portions of parks, or open spaces which are dedicated or recognized by appropriate local officials for activities requiring special qualities of serenity and quiet.
B	70 dBA (exterior)	Residences, motels, hotels, public meeting rooms, schools, churches, libraries, hospitals, picnic areas, playgrounds, active sports areas, and parks.
C	75 dBA (exterior)	Developed lands, properties or activities not included in categories A and B above.
D	55 dBA (interior)	Residences, motels, hotels, public meeting rooms, schools, churches, libraries, hospitals, and auditoriums.

Note: The interior design noise level in Category D applies to indoor activities for those situations where no exterior noise sensitive land use or activity is identified.

Source: From Highway traffic noise: a design guide for highway engineers, Federal Highway Administration, National Cooperative Highway Research Program Report No. 117, 1973. With permission.

The day-night noise level (L_{dn}), which is derived from hourly L_{eq} over a 24 h period, takes into account increased nighttime sensitivity to noise.

As is the case with highways, aviation has its own peculiar set of noise problems. In the case of aviation, however, the noise appears to consist primarily of short-term, high intensity bursts of noise at specific time periods in specified flight paths. Furthermore, the effect of the noise is very specific to the sensitive receptor source. As an example of this, the Federal Aviation Administration (Newman et al., 1982) has developed a set of helicopter noise curves for use in environmental impact assessments. They present the results of FAA measurement programs in 1976, 1978, and 1980 in a single report with data formatted specifically for environmental impact analyses. This data base provides the noise input information necessary to develop helicopter noise contours using a computer model such as the FAA Integrated Noise Model.

Airports have been the subject of a considerable number of noise studies. The Illinois Institute of Natural Resources (Schomer et al., 1981) studied 22 airports (excluding Chicago's O'Hare and Midway Airports) in Illinois. They found that two failed to meet the 1980 limit of 75 dB and an additional 12 would fail to meet the 1985 limit of 65 dB.

10.3   LEGISLATION AND FEDERAL AGENCY POLICIES

The Federal law controlling noise is the Noise Control Act of 1972. Under the Act, the EPA published information on what was known about the levels of noise necessary to protect public health. The noise control part of the EPA was dissolved in the early 1980s and practically all of the noise control activities now reside in the states and municipalities. There are various generally accepted guidelines as to the levels of noise that should be allowed in various types of locations at different hours of the day, and this had led to a patchwork of noise control levels across the country.

The EPA prepared Exhibit 10, which summarizes the yearly equivalent sound levels identified as requisite to protect the public health and welfare with an adequate margin of safety.

The Federal Interagency Committee on Urban Noise (1980) published an excellent description of the noise policies and programs of federal agencies relating to land use. That description is summarized in this section. Following is a list of the agencies involved:

- Department of Defense (DOD).
- Department of Housing and Urban Development (HUD).
- Environmental Protection Agency (EPA).
- Department of Transportation/Federal Aviation Administration (DOT/FAA).
- Department of Transportation/Federal Highway Administration (DOT/FHWA).
- Department of Veterans Affairs (VA).

All of the policies address transportation noise problems, particularly those of highways and airport systems. The policies emphasize these noise sources because federal agencies have provided billions of dollars for their construction and maintenance.

The major differences among the policies center upon the noise levels specified and the types of noise measures used or required. There are four different types of noise levels used in these policies:

- Mitigation levels (e.g., FHWA design levels).
- Levels required to protect the public health and welfare (e.g., the EPA levels document).
- General planning (land use) levels (e.g., DOD).
- Levels required for federal assistance (e.g., HUD, VA), these are similar to the general planning levels.

Because of differences in statutory authority, the noise policies differ in the kinds of noise actions and techniques emphasized. The FAA and EPA regulations stress source and operational controls for aircraft and highway vehicles. The FHWA policy, in the main, stresses noise mitigation at noise sensitive locations along highways. HUD and VA require that the receiver (e.g., residential development) be provided noise attenuation as a condition for mortgage insurance or assistance.

An agency by agency breakdown follows.

Exhibit 10
Yearly Average[a] Equivalent Sound Levels Identified as Requisite to Protect the Public Health with an Adequate Margin of Safety

Measure	Indoor Activity Interference	Indoor Hearing Loss[b]	To Protect Against Both Effects[c]	Outdoor Activity Interference	Outdoor Hearing Loss[b]	To Protect Against Both Effects[c]
Residential with outside space and farm residences						
L_{dn}	45	—	45	55	—	55
$L_{eq}(24)$	—	70	—	—	70	—
Residential with no outside space						
L_{dn}	45	—	45	—	—	—
$L_{eq}(24)$	—	70	70[e]	—	—	—
Commercial						
$L_{eq}(24)$	[d]	70	70[e]	[d]	70	70[e]
Inside transportation						
$L_{eq}(24)$	[d]	70	70[e]	—	70	—
Industrial						
$L_{eq}(24)$[f]	[d]	70	70[e]	[d]	70	70[e]
Hospitals						
L_{dn}	45	—	45	55	—	55
$L_{eq}(24)$	—	70	—	—	70	—
Educational						
$L_{eq}(24)$	45	—	45	55	—	55
$L_{eq}(24)$[f]	—	70	70[e]	—	70	70[e]
Recreational areas						
$L_{eq}(24)$	[d]	70	70[e]	[d]	70	70[e]
Farm land and general unpopulated land						
$L_{eq}(24)$	—	—	—	[d]	70	70[e]

[a] Refers to energy rather than arithmetic averages.

[b] The exposure period that results in hearing loss at the identified level is 40 years.

[c] Based on the lowest level.

[d] Because different types of activities appear to be associated with different levels, identification of a maximum level for activity interference may be difficult except in those circumstances where speech communication is a critical activity.

[e] Based only on hearing loss.

[f] An $L_{eq}(8)$ of 75 dB may be identified in these situations so long as the exposure over the remaining 16 h per day is low enough to result in a negligible contribution to the 24 hr average, that is, no greater than an L_{eq} of 60 dB.

Source: Information on levels of environmental noise requisite to protect public health and welfare with an adequate margin of safety; U.S. Environmental Protection Agency, Publication 550/9-74-004, Office of Noise Abatement and Control, Washington, D.C., 1974. With permission.

10.3.1 DEPARTMENT OF DEFENSE (DOD)

The Department of Defense policy for noise compatible land-use guidance is called the air installation compatible use zone (AICUZ). Each military service studies noise exposure and land use at all DOD air installations. Each study contains noise contours, accident potential zones, existing and future land-use compatibilities and incompatibilities, and land-use planning/control recommendations. Department of Defense policy requires that all reasonable, economical, and practical measures be taken to reduce and/or control the generation of noise from flying.

10.3.2 DEPARTMENT OF HOUSING AND URBAN DEVELOPMENT (HUD)

The major purpose of the Department of Housing and Urban Development's (HUD) noise regulations (24 CFR Part 51 Subpart B) is to ensure that activities assisted or insured by the department achieve the goal of a suitable living environment. The regulations apply to all HUD actions and provide minimum national standards to protect citizens against excessive noise in their communities and places of residence. HUD assistance for construction of new noise sensitive uses is prohibited for projects with unacceptable noise exposures and is discouraged for projects with normally unacceptable noise exposures. Unacceptable noise exposure is defined as a noise level above 75 dB [day–night average sound level (DNL) in decibels]. A normally unacceptable level is one above 65 dB but not exceeding 75 dB. These noise levels are to be based on noise from all sources, highway, railroad, and aircraft. Attenuation measures are required before projects in the normally unacceptable zone can be approved.

10.3.3 ENVIRONMENTAL PROTECTION AGENCY (EPA)

The key statutory mandates under which EPA operated until its noise group was abolished are the Noise Control Act of 1972 (PL92-574) and the Quiet Communities Act of 1978 (95-609). The EPA set noise source emission standards for various products, including transportation vehicles, construction equipment, and consumer products. The EPA also proposed aircraft/airport regulations to the FAA following a special procedure specified in the Noise Control Act of 1972. The EPA Levels Document established threshold levels of impact which, if met, would protect the public with an adequate margin of safety.

In a grants program under the Quiet Communities Act, the EPA initiated such technical assistance programs as the Quiet Communities Program (QCP) and Each Community Helps Others (ECHO).

10.3.4 DEPARTMENT OF TRANSPORTATION/FEDERAL AVIATION ADMINISTRATION (DOT/FFA)

The Federal Aviation Administration's noise program is guided by the 1976 Aviation Noise Abatement Policy and the Aviation Safety and Noise Abatement Act of 1979. The FAA uses two major approaches to implement this policy. The first includes a program to retrofit engines or equipment on noisy aircraft or to replace them with newer, quieter aircraft. It also includes the development of operational procedures which can reduce the aircraft's noise impacts.

The other major approach to noise compatibility is through planning and development activities at airports under the Airport and Airway Development Act of 1970 (as amended). Airport noise control and land-use compatibility (ANCLUC) planning studies integrate the master planning study activities, the environmental considerations, and the airport–land-use compatibility planning activities at an airport. The objective is to achieve maximum noise and environmental compatibility within the constraints of safety, service, and economic viability.

10.3.5   DEPARTMENT OF TRANSPORTATION/FEDERAL HIGHWAY ADMINISTRATION (DOT/FHWA)

As a result of the Federal Aid Highway Act, the Federal Highway Administration (FHWA) is concerned with traffic and construction noise associated with federal aided highways. It requires study of future noise exposures in conjunction with standards featuring highway design noise levels. The FHWA also provides for noise mitigation on existing federal aided highways. For each new highway, the FHWA requires that state highway agencies furnish localities with information on noise and land use.

10.3.6   DEPARTMENT OF VETERANS AFFAIRS (VA)

The Department of Veterans Affairs (VA) policy for consideration of noise and land-use planning is promoted in the VA's Loan Guaranty Program, the Department of Medicine and Surgery (DM&S), and the Department of Memorial Affairs (DMA).

The VA Loan Guaranty noise policy governs VA decisions as to whether residential sites in airport environs are acceptable for loan guaranty programs to eligible veterans and active duty personnel. It sets three noise zones. In the case of new construction, all new developments located in the two higher zones generally are not eligible for VA assistance. The policy for land acquisition and maintenance adhered to by the DM&S and DMA considers noise in the environmental planning of all acquisition and construction programs.

Exhibit 11 (Federal Interagency, 1980) sums up the federal agency noise policies described previously.

10.3.7   EIS INFORMATION ON EXISTING NOISE

Information that should be included in an EIS on the existing situation with regard to noise is as follows:

- The existing and anticipated land uses at and near the project site.
- The existence of sensitive receptors near the project site at which noise measurements may be made.
- Applicable noise standards and criteria for the area: these are usually local, but, on occasion, may be state-wide, for example, for highways.

Exhibit 11
Federal Agency Policy and Program Summary

Agency	Department of Defense (DOD)	Department of Housing and Urban Development (HUD)	Environmental Protection Agency (EPA)	DOT/Federal Aviation Administration (FAA)	DOT/Federal Highway Administration (FHWA)	Veteran's Administration (VA)
Type of program or policy	Air installations compatible use zones (AICUZ) program	HUD noise regulations	Health and welfare guidance	Aviation noise abatement policy	Highway noise policy	VA noise policy
Key documents	DOD Instruction 4165.57 (1977) Installation AICUZ studies	24 CFR Part 51 Subpart B; Noise Assessment Guidelines (1980)	EPA levels Document (1974)	DOT/FAA Aviation Noise Abatement Policy (1976) Advisory Circular: 150/5050-6 (1977)	FHPM 7-7-3 (1976)	Section VIII Appraisal of Residential Properties Near Airports (1969)
Type of noise levels	Levels used as reasonable guidance to communities in planning	Levels which determine whether proposed sites are eligible for HUD insurance or assistance	Levels which are required to protect the public health and welfare with an adequate margin of safety	Levels used as starting points in determining noise/land-use relationships	Design noise levels	Levels determining whether proposed sites are eligible for VA assistance
Purpose of noise levels	Guidance to communities for planning. Reflects cost, feasibility, past community experience, general	See above. Levels can be used as general planning levels. Reflect costs, feasibility, general program objectives, and consideration of health	These levels identify in scientific terms the threshold of effect. While the levels have relevance for planning, they do	Guidance to communities for planning. Reflects safety, cost, feasibility, general program	These levels are used in determining where noise mitigation on a particular highway project is warranted. They do reflect cost	See above. Reflects cost, feasibility, general program objectives and consideration of health and

	Military airfields	All sources	All sources	Civil airports	Highways only	Airports only
	program objectives, and consideration of health and welfare goals.	and welfare goals.	not in themselves form the sole basis for appropriate land-use actions because they do not consider cost, feasibility, or the development needs of the community. The user should make such tradeoffs.	objectives, and consideration of health and welfare goals.	and feasibility considerations. They are not appropriate land-use criteria. Location specific.	welfare goals.
Source to which noise levels applied						

Source: Guidelines for considering noise in land use planning control, Federal Interagency Committee on Urban Noise, 1980. With permission.

- Existing noise levels at the project and the sensitive receptors.
- L_{10}, L_{50}, and L_{90} levels should be given in dBA units. If the noise contains strong low frequency components, dBC scale measurements should also be made.

Measurement of noise levels is a simple procedure that uses a hand-held instrument. The sound level meter is a battery operated device that contains the appropriate electronics to convert the sound pressure exciting the microphone diaphragm into a meter reading in decibels (dB). The concern with using a portable sound level meter is the possible effect the person holding the meter has on the sound field. Certain combinations of noise spectra and distances of the sound level meter from one's body can alter the sound field by several dB. For precision measurements, these effects can be eliminated by simply placing a 2 to 3 meter cable between the sound level meter and the microphone, and mounting the microphone on a tripod.

For many measurements, it is sufficient to simply hand-hold the sound level meter and record the levels directly from the meter reading. However, under certain circumstances it is necessary for the sound level meter to have additional capabilities. To record the noise using a portable tape recorder, the sound level meter must have an AC output terminal. In addition, in order to calibrate the entire sound recording system, a calibrator or pistonphone must be used.

For measuring vehicular pass-by levels or one-time, short duration (impulsive type) noise events, a hold circuit is a necessity for two reasons:

1. Because the duration of the peak pass-by or impulse noise is often less than or equal to the response time of the meter, the hold circuit electronically holds the maximum value, permitting the meter itself sufficient time to respond.
2. Because the peak sound-pressure levels only last a brief moment, it would be difficult to accurately and consistently read the meter (if it could respond correctly) in such a short period of time.

The next step is the prediction of the noise impacts. Noise levels, for the alternatives under consideration during both the construction and operational phases, are identified. A determination is made of the microscale impact by predicting anticipated noise levels for each alternative during both construction and operational phases. Predicted noise levels are compared with applicable standards or criteria in order to assess impact.

The primary noise impacts of construction and plant operation (wastewater and industrial) are determined by calculation. Secondary noise impacts associated with population growth are calculated for residential areas primarily exposed to noise generated by sources other than airports or freeways. The calculated construction, operation, and induced population growth associated noise levels are compared with the measured existing sound levels to assess noise impacts. A comparison of predicted noise levels is made with applicable state and local noise regulations, as well as EPA guidelines to protect the public health and safety.

In the case of highways, the Federal Highway Administration of the U.S. Department of Transportation has developed three procedures whereby traffic noise from freely flowing highway traffic can be reasonably well-predicted. (Federal Highway Administration, 1973) Two of these procedures are graphical and the third requires a digital computer program. One of the graphical procedures uses a readily available nomograph, which is valid for moderately high volume, freely flowing traffic on infinitely long, unshielded, straight, level roadways. Adjustments are then made to the values obtained from the nomograph to include some of the effects of roadway geometry and road surface characteristics.

There are many situations where the traffic flow is intermittent, where cars and trucks operate in accelerating and decelerating modes, or where the principal sound source is an intermittent line of low speed, low volume trucks climbing a steep ramp grade. Simple and reliable noise prediction schemes for such complicated situations are not yet available.

The impacts of aviation traffic on humans have been discussed at length earlier in the first part of this chapter. They include such factors as reduced property values to owners and occupants of nearby land, difficulty in speech communication, annoyance and interruption of work, learning, sleep, and recreation.

10.4 MITIGATIVE MEASURES

Mitigating measures are available for excessive noise both in the construction and operational phases. Noise control measures during the construction phase generally consist of placing muffling devices or other devices which will reduce vibration, scheduling noise generation during the early morning and late evening hours, and enclosing the noise source or absorbing the noise waves on the equipment that produces the noise. There should be an equipment roster with associated sound levels for each piece of equipment and the expected duration of each phase of the construction. In some cases, where a sensitive receptor area is located close to the concentration, a noise berm may be built as a sound deadener. This has been used successfully in cases of noise generation adjacent to zoos. Speed of construction vehicles is kept low.

The noise control measures during operation of a facility are tailored to the source of the noise. In general, noise reduction in industrial plant planning basically consists of the use of quieter equipment, processes, and materials, as well as the reduction of transmission of noise by air through enclosure of the source of the noise. In addition, absorption of the noise by sound proof materials often is practiced.

Noise abatement measures for highways have been summed up by the Federal Highway Administration (1980 and 1990) as follows:

1. *Traffic management measures.* In urban settings, this can be done by approaches such as rerouting trucks so that they take acceptable alternative routes and are prohibited from certain streets and roads, or they can be permitted to use certain streets and roads only during daylight hours. For cars, traffic lights can be changed to smooth out the flow of traffic and to eliminate the need for frequent stops and starts which generate noise.

Speed limits can be reduced; however, about a 20 mile per hour reduction in speed usually is necessary for a noticeable decrease in noise levels. In addition, of course, the removal of drivers from the road and placing them into mass transit modes is very effective. This can be combined with parking controls that encourage the use of mass transit.

2. *Buffer zones, or land use control.* Out in the suburban or rural areas, buffer zones of undeveloped, open spaces which border a highway can be created. This happens when a highway agency purchases land, or development rights, in addition to the normal right-of-way, so that future dwellings cannot be constructed close to the highway. This prevents the possibility of constructing dwellings that would otherwise experience an excessive noise level from nearby highway traffic. An additional benefit of buffer zones is that they often improve the roadside appearance. However, because of the considerable amount of land required, creating buffer zones is often not possible. Where highways are constructed in undeveloped land, it may be possible to retain that land as such.

3. *Planting vegetation.* About 10 years ago, this was the favored approach to highway noise mitigation. Wild rose buses and evergreens sprouted along the sides and median strips of most highways. This is because vegetation, if it is high enough, wide enough, and dense enough that it cannot be seen through, can decrease highway traffic noise. A 200 ft width of dense vegetation can reduce noise by 10 dB, which in practice, cuts in half the loudness of traffic noise. It is often impractical, however, to plant enough vegetation along a road to achieve such reductions.

4. *Insulating buildings.* Insulating buildings alongside the highway can greatly reduce highway traffic noise, especially when windows are sealed and cracks and other openings are filled. Sometimes noise-absorbing material can be placed in the walls of new buildings during construction. However, this type of insulation can be costly. In addition, as is often the case, solving one environmental problem may create another. In this instance, tightly sealed buildings tend to cause degradation of indoor air quality, concentrating such hazards as radon, urea-formaldehyde decomposition products, volatile solvents, and so on.

5. *Highway relocation.* This alternative raises the possibility of altering the highway location to avoid those land-use areas which have been determined to have a potential noise impact. It may also be possible to obtain noise mitigation by elevating or depressing the roadway to produce a break in the line of sight from the source to the receiver.

6. *Noise barriers.* The most popular and frequently used noise control system for highways these days consists of the erection of noise barriers along the side of the road. Noise barriers are solid obstructions built between the highway and the buildings along the highway. Effective noise barriers can reduce noise levels by 1 to 15 dB, cutting the loudness of traffic noise in half. Barriers can be formed from earth mounds along the road (usually called earth berms) or from high, vertical walls. Earth berms have a very

natural appearance and are usually attractive. However, they can require much land. Walls take less space. They are usually limited to 25 ft in height because of structural and aesthetic reasons. Noise walls can be built out of wood, stucco, concrete, masonry, metal, and other materials. Noise barriers can be visually pleasing and blend in with their surroundings.

Exhibit 12 shows the information available from the U.S. Department of Transportation (1998) on highway noise barrier construction in term of length (miles) and costs from 1970 to 1995.

It should be noted that noise barriers are not always effective. The barrier must be high enough both to block the view of a road and also to prevent sound waves from reaching the upper levels of nearby houses. This limits the utility of barriers for those locations where the homes are on a hillside overlooking the road. Openings in noise walls for driveway connections or intersecting streets destroy the effectiveness of barriers. In some areas, homes are scattered too far apart to permit noise barriers to be built at a reasonable cost. Overall, noise barriers appear to have been accepted by the general public in a positive way. Residents adjacent to barriers have stated that conversations in households are easier, sleeping conditions are better, a more relaxing environment is created, windows are opened more often, and yards are used more in the summer. Perceived nonnoise benefits included privacy, cleaner air, improved view and sense of ruralness, and healthier lawns and shrubs. Negative reactions have included a restriction of view, a feeling of confinement, a loss of air circulation, a loss of sunlight and lighting, and poor maintenance of the barrier. Most residents near a barrier seem to feel that barriers effectively reduce traffic noise and that the benefits of barriers outweigh the disadvantages of the barriers (FHWA, 1990).

Aircraft noise abatement involves modification of aircraft design, changes in aircraft operations and route locations, more frequent aircraft maintenance, and landscape architecture and acoustic insulation at facilities located on or near airports (DOT, 1972). In addition, land-use zoning in the vicinity of airports may be practiced (Canter, 1977).

The level of noise at the nation's airports and surrounding areas is declining as airlines take older, noisier airplanes out of service and replace them with newer, quieter ones. The Airport Noise and Capacity Act of 1990 requires that all airplanes meet quieter Stage 3 noise levels by the year 2000.

A report recently submitted to Congress by the DOT shows that operators were ahead of interim compliance requirements to either reduce the number of noisier Stage 2 airplanes by 50 percent or have 65 percent quieter Stage 3 airplanes in their fleets. As of December 1997, 75.5 percent of the airplanes operating in the United States are Stage 3.

"Aviation operators are continuing to be good neighbors to communities impacted by noisier aircraft noise", said FAA Administrator Jane F. Garvey. "Just this past year, 370 noisier Stage 2 aircraft have been removed from service while 230 quieter Stage 3 aircraft have entered service in the United States."

Stage 2 airplanes include Boeing models 727-200, 737-200 and McDonnell Douglas model DC-9. Stage 3 airplanes include Boeing models 737-300, 757, 777, and McDonnell Douglas model MD-90. Some operators are complying with the

Exhibit 12
Highway Noise Barrier Construction (Miles)

	Unknown	1970–79	1980–89	1990	1991	1992	1993	1994	1995	1970–95
Length, Total	6	177	567	65	101	141	65	61	115	1,318
Type I Barriers[a]	6	104	422	45	79	112	60	41	78	947
Type II Barriers[b]	0	71	130	20	20	19	22	16	31	329
All Other Types[c]	N	2	15	1	2	10	3	4	6	43
Cost (Millions of 1995 Dollars)	N	130	624	92	142	184	112	72	141	1,497

[a] A Type I barrier is built on a highway project to construct a new highway or to physically after an existing highway.

[b] A Type II barrier is built to abate noise along an existing highway (often referred to as retrofit abatement), and is not mandatory.

[c] All other types of barriers are nonfederally funded.

N Data are nonexistent.

Stage 2 airplane phaseout by installing FAA certified Stage 3 noise level hushkits to their Stage 2 fleet.

The Helicopter Association International (1983) has published a guide that describes a voluntary noise reduction program designed to be implemented worldwide by all types of civil, military, and governmental helicopter operations. General procedures are recommended to minimize acoustical impacts. Specific procedures are described for the approach, take-off, and enroute situations. The abatement procedures have to do with height, side of the helicopter facing the noise sensitive area, controls, rate of descent, and so on.

REFERENCES

Aircraft noise levels continue to decline, *FAA News,* Washington, D.C., 1997.

Bond, J., Winchester, C. F., Campbell, L. E., and Webb, J. C., Effects of Loud Sound on the Physiology and Behavior of Swine, U.S. Department of Agriculture, Agricultural Research Service Technical Bulletin No. 1280, 1980.

Canter, L. W., *Environmental Impact Assessment,* McGraw-Hill, New York, 1977.

Edwards, R. G., Broderson, A.B., Barbour, R. W., et al., Assessment of the Environmental Compatibility of Differing Helicopter Noise Certification Standards, U.S. Department of Transportation, Federal Aviation Administration, 1979.

Fly Neighborly Guide, Helicopter Association International, ISSN 0739-8581, 1983.

Highway Traffic Noise, HEV-21/8-80 (20-M), Federal Highway Admimistration, SEA, 1990.

Guidelines for Considering Noise in Land Use Planning Control, Federal Interagency Committee on Urban Noise, 1980.

Highway noise barrier construction, U.S. Department of Transportation, web site, http://www.bts.gov/bcsprod/nts/chp4/tbl4x45.html, 1998.

Highway Traffic Noise, HEV-21/8-80(20M), Federal Highway Administration, 1980.

Highway Traffic Noise: A Design Guide for Highway Engineers, National Cooperative Highway Research Program Report No. 117, Federal Highway Administration, 1973.

Information on Levels of Environmental Noise Requisite to Protect Public Health and Welfare with an Adequate Margin of Safety, Environmental Protection Agency, Publ. 550/9-74-004, Office of Noise Abatement and Control, Washington, D.C., 1974

Newman, J. S. and Beattie, K. R., Aviation Noise Effects, U.S. Department of Transportation, Report No. FAA-EE-85-2, 1985.

Newman, J. S., Rickley, E. J., and Bland, T. L., Helicopter noise exposure curves for use in environmental impact assessment, 1982.

Parker, J. B. And Bayley, N. D., Investigations on Effects of Aircraft Sound on Milk Production of Dairy Cattle, U.S. Department of Agriculture, Agricultural Research Service, Animal Husbandry Research Division, 1960.

Schomer, P., Findley, R. W., and Frankel, M., *Economic Impact of Proposed Airport Noise Regulations,* R77-L, Vol. I, Illinois Institute of Natural Resources Document No. 81/02, 1981.

Transportation Noise and Its Control, U.S. Department of Transportation, Publication DOT P 5630.1, 1972.

11 The Man-Made Environment: Hazards and Nuisances

The material that belongs in this portion of an EIS covers two areas as follows:

- Conventional hazards and nuisances such as high voltage power lines, buried oil and gas lines, and so on.
- Toxic and hazardous waste sites, underground storage tanks and other sources of contamination of land and groundwater that are regulated by the federal government.

The discussion that follows includes both of the above items. A discussion of how RCRA and Superfund activities are covered by a NEPA substitute follows. Finally, a brief review is made of how these items should be handled in a NEPA study.

11.1 CONVENTIONAL HAZARDS AND WASTES

Hazards and nuisances at and in the vicinity of the proposed sites should be identified. Man-made hazards and nuisances in the study areas, such as major natural gas and petroleum pipelines, odors, high wind velocities, and high voltage transmission lines, are described. The impacts of these hazards and nuisances on the proposed projects are determined. These types of hazards and nuisances are site-specific and may result in a change in project site.

11.2 REGULATED HAZARDOUS WASTES

This section will begin with a generalized discussion of each of the two major federal laws—RCRA and Superfund—because these are the two types of hazardous and solid waste activities that are of the most concern to persons doing a NEPA study. After that discussion, the relations of each to NEPA will be described. Finally, a subsection will be devoted to the inclusion of information about them in a NEPA study.

11.2.1 RESOURCE CONSERVATION AND RECOVERY ACT (RCRA)—GENERAL

RCRA is intended to provide cradle to grave management of hazardous wastes, management of solid wastes, and regulation of underground storage tanks containing chemical and petroleum products.

A waste is considered hazardous if it exhibits hazardous characteristics such as corrosivity, reactivity, ignitability, or extraction procedure toxicity, or if it is specifically listed in a regulation by the EPA. Wastes excluded from regulation as hazardous wastes are household wastes, crop or animal wastes, mining overburden, wastes from processing and beneficiation of ores and minerals, flyash, bottom ash, slag waste, flue gas emission control waste, and drilling fluids from energy development.

Solid wastes, if land disposed, are regulated through state programs under Subtitle D of RCRA. Solid waste is defined in RCRA to include garbage, refuse and sludge, and other solid, liquid, semi-solid, or contained gaseous material that is discarded. Exclusions from solid waste include domestic sewage, irrigation return flow, material defined by the Atomic Energy Act, *in situ* mining waste, and NPDES point source wastes.

Subtitle I enables national regulation of underground storage tanks for the first time. In practice, the individual states presently do most of the regulating. Underground storage tanks containing hazardous wastes are regulated under Subtitle C.

Section 3004 requires the EPA to promulgate standards applicable to transporters of hazardous waste. It requires the transportation of hazardous waste to a treatment, storage, or disposal facility only if the waste is properly labeled and in compliance with a manifest system that provides a permanent record of the waste at all times.

Section 3004 also requires the EPA to promulgate standards applicable to owners and operators of hazardous waste treatment, storage, and disposal facilities. There are several significant provisions to this section, including bans on liquids in landfills, the development of standards for facilities that produce fuel from hazardous waste, and corrective action at permitted facilities and beyond facility boundaries.

Section 3005 provides permit requirements for facilities that treat, store, or dispose of hazardous waste.

More information concerning specific RCRA requirements follows.

11.2.2   LAND DISPOSAL

RCRA prohibits the continued land disposal of untreated hazardous waste beyond specified dates, unless a petitioner demonstrates that the hazardous constituents will not migrate from the land disposal unit for as long as the waste remains hazardous. For purposes of restriction, Congress defined land disposal under RCRA to include any placement of hazardous waste in a landfill, surface impoundment, waste pile, injection well, land treatment facility, salt dome or salt bed formation, or underground mine or cave. An applicant, such as the owner or operator of a treatment, storage, or disposal facility, may petition the EPA to allow land disposal of a specific waste at a specific site. The applicant must prove that the waste can be contained safely in a particular type of disposal unit, so that no migration of any hazardous constituents occurs from the unit for as long as the waste remains hazardous. If the EPA grants the petition, the waste is no longer prohibited from land disposal in that particular type of unit.

Part 262 of RCRA provides standards for generators of hazardous wastes.

Part 263 provides standards applicable to transporters of hazardous waste and Part 264 provides standards for hazardous waste management, storage, and disposal facilities. The latter provides for contingency plans and emergency procedures. There are groundwater protection standards and groundwater monitoring provisions. There are conditions for closure and postclosure care related to the facility, as well as financial assurance for closure operations. Part 264 addresses containers, tanks, surface impoundments, waste piles, land treatment, landfills, and incinerators.

Part 270 addresses treatment, storage, and disposal permits. The permit procedure has two parts. The Part A application requires identification and other general information about the facility including the types of hazardous wastes to be treated, stored, or disposed of and an estimate of the quantity of such wastes to be treated, stored, or disposed of annually. In addition, Part A requires a listing of all permits obtained and a topographic map extending one mile beyond the property boundaries of the source.

Part B of the permit application requires both general and specific information. These requirements include:

1. A general description of the facility.
2. Chemical and physical analyses of the hazardous waste to be handled at the facility.
3. A copy of a waste analysis plan.
4. A description of the security procedures and equipment.
5. A copy of the schedule for facility inspections of malfunctions or discharges to the environment.
6. A copy of the contingency plan in the event of fires, explosions, or any unplanned release of hazardous waste to the air, soil, or surface water.
7. A description of procedures, structures, or equipment used at the facility to prevent hazards in unloading operations, to prevent run-off from hazardous waste handling areas, to prevent contamination of water supplies, to mitigate effects of equipment failure and power outages, and to prevent undue exposure of personnel to hazardous waste.
8. A description of precautions to prevent accidental ignition or reaction of ignitable, reactive, or incompatible wastes.
9. A description of traffic patterns at the facility, estimated volume and control of traffic, and description of access road surfacing and load bearing capacity.
10. Information related to facility location, seismic zones, and 100 year floodplain.
11. An outline of both the introductory and continuing training programs by owners or operators to prepare persons to operate or maintain the facility in a safe manner.
12. A copy of the closure plan and the postclosure plan where applicable.
13. Information on the most recent closure cost estimate for the facility.

14. A topographic map showing a distance of 1000 ft around the facility at a scale of 1 in. to not more than 200 ft. Contours must be shown on the map. The contour interval must be sufficient to clearly show the pattern of surface water flow in the vicinity of and from each operational unit of the facility.

For owners or operators of hazardous waste surface impoundments, waste piles, land treatment units, and landfills, additional information regarding protection of groundwater is required. This information includes:

- A summary of groundwater monitoring data.
- Identification of the uppermost aquifer and aquifers hydraulically interconnected beneath the facility property, including groundwater flow direction and rate, and the basis for such identification.
- Detailed plans and an engineering report describing the proposed groundwater monitoring program to be implemented.
- If the presence of hazardous constituents has been detected in the groundwater at the point of compliance at the time of permit application, the owner or operator must submit sufficient information, supporting data, and analyses to establish a compliance monitoring program, and submit an engineering feasibility plan for a corrective action program.

In a NEPA type study, there may be a facility present that is subject to RCRA. If so, it is important to know whether it is in compliance and, if not, what the shortcomings are. The nature of the operation may have a negative impact upon the project that is being studied in the NEPA process.

11.2.3   UNDERGROUND STORAGE TANKS

The Superfund Amendments and Reauthorization Act of 1986 amended RCRA to provide for further regulation of underground storage tanks (UST) that store useful materials, as well as wastes. The law was passed because there was a growing concern over the increasing number of incidents where gasoline vapors were detected in houses and where drinking water was contaminated by leaking petroleum tanks. The Superfund statute excludes petroleum releases from its jurisdiction. Until this law was passed, there was no way to clean up leaks of petroleum products from underground tanks.

An underground tank is defined as any tank with at least 10 percent of its volume buried below ground, including any pipes attached. Thus, tanks with extensive underground piping may now be regulated. Certain tanks are excluded from the law. These include farm and residential tanks holding less than 1100 gallons of motor fuel, on-site heating oil tanks, septic tanks, systems for collecting storm and wastewater, and liquid traps or gathering lines related to oil and natural gas operations.

The law gives the EPA, and the states that enter into cooperative agreements with the EPA, the authority to issue orders requiring owners and operators of underground storage tanks to undertake corrective actions where a leak is suspected. These corrective actions could include testing tanks to confirm the presence of a leak, excavating the site to determine the exact nature and extent of contamination, and cleaning contaminated soil and water. They may also include providing an alternative water supply to affected residences, or temporary or permanent relocation of residents.

The Congress believes that payment of cleanup costs can be satisfied by pollution liability insurance maintained by tank owners and operators. The EPA is directed by the law to publish regulations requiring all tank owners and operators, including those owning chemical tanks, to maintain the financial capability to clean up leaks. For petroleum production, refining, and marketing facilities, Congress has set minimum coverage levels at $1 million per occurrence.

A regulation banning underground installation of unprotected new tanks went into effect in 1985. A new underground storage tank cannot be installed unless:

1. It will prevent release of the stored substance owing to corrosion or structural failure for the life of the tank.
2. It is protected against corrosion, constructed of noncorrosive material, or designed to prevent release of the stored substance.
3. Construction or lining materials are electrolytically compatible with the substance to be stored.

The law specifies that leak detection, prevention, and corrective action regulations must require owners and operators of underground storage tanks to:

1. Be able to detect releases.
2. Keep records of release detection methods.
3. Take corrective action when leaks occur.
4. Report leaks and corrective action.
5. Provide for proper tank closure.
6. Provide evidence, as the EPA deems necessary, of financial capability to take corrective action and compensate third parties for injury or damages from instant or continuous releases. States may finance corrective action and compensation programs by a fee levied on owners and operators.

The same requirements are being extended to all existing UST as of the date that this book is being published.

11.2.4 SUPERFUND (CERCLA)

Hazardous waste are produced in the United States at the rate of 700,000 tons per day, or over 250 million tons per year (EPA, 1987). Because of the uncontrolled deposition of hazardous wastes in the past, thousands of abandoned or inactive sites containing hazardous wastes have been identified nationwide. Many of these sites are located in environmentally sensitive areas, such as floodplains or wetlands. Rain and melting snow seep through the sites, carrying chemicals that contaminate underground waters and nearby streams and lakes. At some sites, the air also is contaminated as toxic vapors rise from evaporating liquid wastes or from uncontrolled chemical reactions.

Superfund, the Comprehensive Environmental Response, Compensation and Liability Act of 1980, was created to cleanup the hazardous waste mistakes of the past and to cope with the emergencies of the present. The objectives of Superfund are to develop a comprehensive program to set priorities for cleaning up the worst existing hazardous waste abandoned or uncontrolled sites; to make responsible parties pay for cleanup wherever possible; and to operate under a trust fund for the purposes of performing remedial cleanups in cases where responsible parties cannot be held accountable, as well as responding to emergency situations involving hazardous substances.

Many Superfund sites were created by the chemical and petroleum industries. Others were once municipal landfills that have become hazardous as a result of accumulated pesticides, cleaning solvents, and other chemical products discarded in household trash. Many sites are the result of transportation spills or other accidents, and others are the final resting place of persistent toxic pollutants contained in industrial wastewater discharges or air pollution emissions (EPA, 1987).

The EPA has established a national priorities list (NPL), which is a list of priority sites for long-term remedial response and cleanup. Only those sites included on the national priorities list (NPL) are eligible for financial remedial action with funds supplied by the trust fund. Sites are nominated by the EPA for the NPL as a result of a hazard ranking system (HRS) that evaluates the threat a site poses to human health or to the environment. In addition, each state or territory may designate one top-priority site, regardless of score. According to the EPA, by March 1998, of the 57 states and territories, 40 had designated top-priority sites, of which 7 had already been deleted from the NPL because no further action was necessary (EPA, 1998).

A third and infrequently used approach in placing sites on the NPL has been if the site meets three requirements:

- A recommendation by the Agency for Toxic Substances and Disease Registry of the U.S. Public Health Service that people be moved from the site.
- A determination by the EPA that the site poses a significant threat to public health.
- Anticipation by the EPA that it is more cost-effective to use its remedial authority than its emergency response authority for the site.

Under these 3 provisions, 13 sites have been listed. The NPL as of March 1998 had 1197 final sites and 54 proposed sites (EPA, 1998) including some federal facilities.

Exhibit 13 breaks down the NPL by states. As of 1995 (EPA, 1996), those parties responsible for contamination had performed 75 percent of new Superfund cleanups and, since 1986, had committed to pay more than $11 billion towards those cleanups. In 1995 alone, over $670 million was spent cleaning up hazardous waste.

For those cases where the responsible party is as yet unknown, the EPA cleans up the site using funds from the Superfund Trust Fund that comes from chemical,

EXHIBIT 13
National Priorities List

States	NPL Totals	States	NPL Totals	Totals
Alabama	13	Nebraska	10	
Alaska	7	Nevada	1	
Arizona	10	New Hampshire	18	
Arkansas	11	New Jersey	110	
California	94	New Mexico	10	
Colorado	17	New York	80	
Connecticut	14	North Carolina	23	
District of Columbia	1	North Dakota	0	
Delaware	17	Ohio	37	
Florida	55	Oklahoma	11	
Georgia	17	Oregon	11	
Hawaii	4	Pennsylvania	100	
Idaho	9	Puerto Rico	10	
Illinois	41	Rhode Island	12	
Indiana	30	South Carolina	26	
Iowa	17	South Dakota	2	
Kansas	11	Tennessee	15	
Kentucky	16	Texas	30	
Louisiana	16	Utah	16	
Maine	12	Vermont	8	
Maryland	16	Virginia	26	
Massachusetts	31	Virgin Islands	2	
Michigan	74	Washington	47	
Minnesota	28	West Virginia	7	
Mississippi	3	Wisconsin	39	
Missouri	22	Wyoming	3	
Montana	9			
				1251

Source: From U.S. Environmental Protection Agency, Office of Solid Waste and Emergency Response, Publication 9320.7-061, March 1998. With permission.

petroleum, and corporate taxes. The states where the sites are located must pay at least 10 percent of the cleanup costs and are responsible for the operation and maintenance of those sites. Although the EPA lays out the cleanup costs, it continues its search for potential responsible parties (PRPs). Once the PRPs are located, the EPA sends them notice letters. A notice letter summarizes the information the EPA has used to identify the PRPs and encourages them to work with the EPA to agree on cleanup responsibilities for the site.

PRPs may be responsible for the entire cost of the cleanup; therefore, the early negotiation of a fair cleanup plan with the EPA will save them time and money in the long run. If the PRPs do not cooperate, the EPA can either get a court order requiring them to perform the cleanup or continue to conduct the cleanup itself using the Trust

Fund. If the EPA conducts the cleanup, the agency then can recover in court up to three times the amount of the cost of the cleanup plus penalties. The Trust Fund also pays for cleanup if the PRPs cannot be found or if they are unable or unwilling to pay.

It is illegal for any person to knowingly fail to notify the EPA of the existence of any hazardous waste facility for hazardous waste disposal. The 1986 Amendments authorized increased criminal penalties for failure to report releases of hazardous waste and made the providing of false or misleading information a criminal offense. Statutory authority was given to the use of settlement agreements and the establishment of specific procedures for reaching them. The powers of EPA access to hazardous waste sites for the completion of investigations and cleanup were increased.

State involvement is a requirement of the 1986 amendments. The EPA must ensure that states participate in identifying National Priorities List sites; the review of all preliminary documents related to Superfund remedial actions, as well as final plans for the actions; all enforcement negotiations and concurrences in settlement agreements; and the deletion of sites from the NPL, such as agreement with the EPA and responsible parties that a Superfund cleanup is complete.

11.3   EMERGENCY PLANNING AND COMMUNITY RIGHT-TO-KNOW

The Emergency Planning and Community Right-to-Know Act of 1986 is Title III of the Superfund Amendments and Reauthorization Act of 1986. Title III requires federal, state and local governments and industry to work together in developing emergency plans and reporting on hazardous chemicals. These requirements build upon the EPA's Chemical Emergency Preparedness Program and numerous state and local programs aimed at helping communities deal with potential chemical emergencies. The community right-to-know provisions allow the public to obtain information about the presence of hazardous chemicals in their communities and releases of these chemicals into the environment. These provisions may be important in the case of projects that will have a potential for such chemical emergencies.

Title III has four major sections: emergency planning, emergency notification, community right-to-know reporting requirements, and toxic chemical release reporting.

The emergency planning section is designed to help state and local governments develop emergency response and preparedness capabilities through better coordination and planning, especially within the local community. It requires the governor of each state to designate a state emergency response commission. This state commission should represent state organizations and agencies with expertise in emergency response, such as state environmental, emergency management, and public health agencies.

Private sector groups and associations may also be included. The state commission must designate local emergency planning districts and appoint local emergency planning committees for the districts. The local emergency planning committees must include elected state and local officials; police, fire, civil defense, public health professionals; environmental, hospital, and transportation officials; community groups; and the media.

The state commission supervises and coordinates the activities of the local emergency planning committees, establishes procedures on how to handle requests for information, and reviews local emergency plans. The local emergency planning committee has primary responsibility in developing a plan that will:

- Identify facilities and transportation routes of extremely hazardous substances.
- Describe emergency response procedures.
- Designate a community coordinator and facility coordinator to implement the plan.
- Outline emergency notification procedures.
- Describe community and industry emergency equipment and facilities, and who is responsible for them.
- Describe and schedule a training program to teach methods for responding to chemical emergencies.
- Establish methods and schedules for exercises to test emergency response plans.

Emergency notification requires that facilities where a listed hazardous substance is produced, used, or stored must immediately notify the local emergency planning committee and the state emergency response commission if there is a release of any such substance to the environment. The substances are those on the list of 360 extremely hazardous substances as published in the Federal Register (40 CFR 355) or on a list of 725 substances subject to the emergency notification requirements under CERCLA Section 103(a), 40 CFR 302.4. Some chemicals are common to both lists.

Under the community right-to-know reporting requirements, facilities required to prepare or have available material safety data sheets (MSDS) under the regulations of OSHA must submit copies of them or a list of MSDS chemicals to the local emergency planning committee, the state emergency response commission, and the local fire department. In addition, the facility must submit an emergency and hazardous chemical inventory form to the same groups. The hazardous chemicals are the same as those for which facilities are required to submit MSDS or a list of MSDS chemicals under the first reporting requirement. The inventory form must record:

- An estimate of the maximum amount of covered chemicals present at the facility at any time during the preceding calendar year.
- An estimate of the average daily amount of covered chemicals present.
- The general location of covered hazardous chemicals.

The toxic chemical release reporting affects owners and operators of facilities that have 10 or more full-time employees, that are in Standard Industries Classification Codes 20 through 39 (which include basically all manufacturing industries), and that manufacture, process, or otherwise use a listed toxic chemical in excess of specified threshold quantities. The toxic chemical release form must be submitted to the EPA, as well as to state officials designated by each governor. Submission is on an annual basis. The report contains information on whether a chemical is manufactured, processed, or otherwise used; estimates of the maximum amounts of the toxic chemical present at the facility at any time during the preceding year; waste treatment and disposal methods for dealing with the chemical and the efficiency of the methods for each waste stream; and the quantity of the chemical entering the environment annually.

11.4   FUNCTIONALLY EQUIVALENT ENVIRONMENTAL EISs

NEPA requires the preparation of an EIS for major federal actions about to be undertaken that may result in significant environmental impacts. Nothing in RCRA or CERCLA precludes compliance with the full provisions of NEPA. However, CEQ and the EPA have determined that adherence to the procedural mandate of CERCLA/SARA and RCRA is to be treated as the functional equivalent of an EIS, providing adequate assurances that all environmental factors are given appropriate consideration.

11.5   CERCLA

Owing to the immediate response nature of many Superfund activities, the procedure for complying with the general intent of NEPA presents a greater problem than for other EPA programs. The issue of how to fashion a program approach that meets quick response requirements, while at the same time permitting adequate environmental review of major federal actions, was the subject of much consideration by both Congress and the EPA during the legislative and regulatory process. Following initiation of activities under CERCLA, numerous court decisions have been rendered that address this problem. The courts have upheld EPA's use of the *functional equivalent* of a NEPA review in its permitting and regulatory activities. Given this exemption from requirements for compliance with formal EIS procedures, the EPA is nonetheless held responsible for ensuring that activities under CERCLA are carried out with a full and adequate consideration of environmental issues and alternatives, and for providing the opportunity for public participation and comment before final decisions are made.

In the cases of remedial actions related to Sections 104 and 106 of CERCLA, EPA headquarters has outlined two specific actions to ensure that the criteria for functional equivalence with NEPA are indeed met. First, the process outlined under CERCLA Section 105(3) and 300.68 of the National Contingency Plan allows for the appropriate review of environmental factors and alternatives. Second, the Superfund community relations program developed after passage of CERCLA, and the emphasis on community technical assistance grants and overall technical assistance to communities required under SARA, further strengthen the opportunity for environmental review and public participation.

11.6   RCRA

As with CERCLA, the RCRA program applies the functional equivalent in meeting NEPA objectives. In fact, staff in the EPA's Office of General Council (OGC) have stated that the permit process in RCRA is so detailed and comprehensive and requires such a review of environmental factors, that it is, in fact, equivalent to the requirements of an EIS.

In November 1984, Congress passed the Hazardous and Solid Waste Amendments (HSWA) to the 1976 Resource Conservation and Recovery Act (RCRA), 42 U.S.C. Sections 6901 et seq. (PL 98-616). One provision of these amendments, Section 3004(u), addressed corrective action for continuing releases from hazardous waste treatment, storage, or disposal facilities. Under this provision, a facility applying for a RCRA permit may be subject to a preliminary environmental assessment by the regulatory agency (either the EPA or an authorized state agency) to whom the application has been submitted. If a waste management unit at a facility is suspected to be the source of a contaminant released to the environment, the owner or operator of the facility may be required to perform a RCRA facility investigation (RFI) to define the nature and extent of the release. This information is then used to determine the need for corrective measures and to aid in their formation and implementation.

Adequate public involvement also is required as part of the RCRA permitting process and related procedures, for example, the RCRA information included in the compliance docket. Information submitted to the EPA under RCRA Section 3005 (Permits for Treatment, Storage, and Disposal of Hazardous Waste), Section 310 (Preliminary Notification), and Section 3016 (Inventory of Federal Agency Hazardous Waste Facilities), and on reportable releases under CERCLA Section 103 will be included in the new Federal Agency Hazardous Waste Compliance Docket, which will be made available for public inspection.

In summary, both CERCLA/SARA and RCRA meet the test of functional equivalence. By the nature of the detail and process for assessing environmental factors and raising alternatives that are inherent in both RCRA's permitting process and for Superfund required through the NCP, both programs direct the EPA in the direction of the objectives to be met by NEPA.

11.7   NEPA CONTENTS

The following approach is usually taken for hazards and nuisances in the preparation of a NEPA document. Conventional hazards and nuisances are described in terms of their types and locations. Possible impacts on the project then are considered. For regulated hazardous wastes, the presence or absence of facilities involving them is discussed. Requirements that must be met are noted. Possible impacts on the proposed project are discussed. As an example of this, we present on the following pages an excerpt from a EIS prepared by BREGMAN & COMPANY for the U.S. Department of Veterans Affairs on the construction of a national cemetery in the northern Illinois area (Bregman, 1991).

"5.2.5 Hazards

Hazards at the sites consist only of possible hazardous wastes at Ft. Sheridan. There are no gas or petroleum problems at any of the sites, nor are there any high voltage lines on the sites. A three-phase power line exists near the Grant Park site. This is considered as an asset, as it would allow power to the site.

FT. SHERIDAN SITE

Ft. Sheridan has eight inactive landfills. Most of these were used between World Wars I and II, and have not seen use since. One of the landfills, however, was used into the 1970s; it is located about one mile south of the proposed cemetery site and 0.3 mile north of the south Post gate.

Another three acre landfill is located on the bluff near Lake Michigan in the proposed cemetery area. At one time, this landfill area was an ammunition burning site. During bluff stabilization activities in this area in the mid-1980s, unexploded ammunition was found beneath the ground surface at this site. Appropriate disposal of the ammunition was made and the site was cleaned up in 1985. . . .

Other areas of Ft. Sheridan are being investigated. One such area includes the Nike missile launch pads. During World War II, there was an active shooting range on the Post and the soils in this area will be tested for lead. Areas near the center of the Post that have been associated with maintenance activities also will undergo a testing of soils. The underground fuel tanks associated with the heliport will be removed with closure and prior to any release for use of the area as a national cemetery.

The inactive landfills, Nike missile launch pads, World War II shooting range, and maintenance area soils are being investigated as part of the Ft. Sheridan closure proceedings. This investigation began in October 1989. The investigation is scheduled to be completed in January 1993, and any required site cleanup as a result of findings in the investigation is scheduled to be completed in June 1994. . . .

Ft. Sheridan was placed on the Illinois CERCLIS list as a potential hazardous waste candidate for the National Priority List. Following investigation, the Illinois EPA Land Pollution Office determined that Ft. Sheridan is a No Further Action Site meaning that it will not be further investigated as a Superfund site.

GRANT PARK AND CISSNA PARK SITES

The Grant Park or Cissna Park proposed sites do not appear on the Illinois CERCLIS list, nor is there any indication that hazardous wastes occur on these sites or ever have been placed there. Visual observations confirm this."

REFERENCES

Environmental impact statement for a proposed National Cemetery in northern Illinois, BREGMAN & COMPANY, Inc. for the U.S. Department of Veterans Affairs, Washington, D.C., 1991.

Background information and supplementary materials: National Priorities List, proposed rule and final rule, U.S. Environmental Protection Agency, Office of Solid Waste and Emergency Response, Publication 9320.7-061, March 1998.

Focus on cleanup costs, U.S. Environmental Protection Agency, Office of Solid Waste and Emergency Response, Publication EPA 540-k-96/004, June 1996.

498 construction cleanups, U.S. Environmental Protection Agency, http:www.cpd.gov/superfund/oerr/accomp/400/sites.hem, January 5, 1998.

Superfund: looking back, looking ahead, U.S. Environmental Protection Agency, Office of Public Affairs, Washington, D.C., 1987.

12 The Man-Made Environment: Historic and Cultural Resources

Special attention is given in environmental impact statements to the presence or absence on-site or nearby of historic and cultural resources, any possible impacts upon them and mitigating measures. The term "historic and cultural resources" covers a variety of features, the most significant one of which is archaeology. Others include historic sites (recent history, i.e., up to 200 years old, as distinguished from much older historical sites covered by the term archaeology), architecturally important buildings, locations and facilities that have a cultural significance to the local communities (including those of ethnic, Indian or racial significance), and possibly unique geological locations. The term historic is broad enough to include any sites where history may have been made. Thus, for example, the author of this book oversaw a study in the 1970s that attempted to discover the original site of a Lincoln–Douglas debate. It turned out to be a present-day parking lot.

12.1 LEGAL BACKGROUND

This is a somewhat confusing area, since there are federal, state and local laws that are applicable to historic and cultural resources. Presented below is a brief review of federal requirements followed by a generalized discussion of state and local situations.

The basic act from which NEPA requirements for historic and cultural resources are derived is the National Historic Preservation Act (NHPA) of 1966 (PL89-665), as amended. There are a number of key requirements in this Act and its implementing regulations, the three major ones follow:

1. National Register of Historic Places.
2. Section 106 process.
3. Advisory Council on Historic Preservation.

These three requirements are interwoven and are discussed together below.

The National Register of Historic Places, which predates the Act, is expanded by the Act to include districts, sites, buildings, structures, and objects significant in American history. If an existing adequate survey of archaeological/historic resources at a proposed project site is not available, a new survey may be necessary to identify resources on the site which may potentially be eligible for inclusion in the National Register. Depending on the probability of a project's potential environmental impact, these surveys may include either a reconnaissance (Step I) or intensive (Step II) survey. The surveys are conducted at the early/preliminary design stage of a project. If properties appear eligible, a request for determination of eligibility is forwarded from the lead agency to the State Historic Preservation Officer (SHPO) (36 CFR 63). If both parties agree on the eligibility, a letter stating the agreement, along with a description of the property and the SHPO's statement of eligibility is forwarded by the agency to the Keeper of the National Register. The agency and the SHPO are then notified of the determination of eligibility within ten working days. If the agency and the SHPO disagree on the eligibility, then the Keeper of the National Register is notified and makes a determination within 45 days.

As stated in 36 CFR 800.9, adverse effects on National Register or eligible properties may occur under conditions which include but are not limited to:

1. Destruction or alteration of all or part of a property.
2. Isolation or alteration of the property's surrounding environment.
3. Introduction of visual, audible, or atmospheric elements that are out of character with the property or alter its setting.
4. Neglect of a property resulting in its deterioration or destruction.
5. Transfer or sale of a property without adequate conditions or restrictions regarding preservation, maintenance, or use.

If the property or properties are eligible for inclusion in the Register, then a determination of the effect of a project is made following the Advisory Council Criteria of Effect [36 CFR 800.9(a)]. The five possible adverse effects listed above are considered. If the proponent determines that no effect is anticipated, the project can proceed. The agency is notified and, in turn, notifies the Advisory Council on Historic Preservation which has 15 days to object.

If an effect is found, the applicant, in consultation with the State Historic Preservation Officer (SHPO), must determine whether the effect is adverse according to the Advisory Council Criteria. The Advisory Council has 30 days in which to review the documentation and concur with or reject the finding. If the Advisory Council finds an adverse effect, mitigation may be suggested. If the agency and the applicant agree, conditions to mitigate impacts are included as part of the approval conditions.

If the project is determined to possess adverse impact, the applicant prepares a preliminary case report to develop the mitigating measures. This is forwarded to the Advisory Council. At this point, the project may proceed along either an expedited process or a consultation process. In the case of a noncontroversial project containing impacts that are customarily mitigated in a standard manner, a memorandum of agreement (MOA) signed by the applicant, the agency, and the SHPO may be delivered to the Advisory Council. This memorandum of agreement sets forth measures to minimize potential adverse effects and establishes an agency review process for it. If the project cannot be expedited, the applicant, the SHPO, the agency and the executive director of the Advisory Council must meet to produce an MOA with mitigating measures. If no agreement can be reached, an EIS may be required. The EIS must contain sufficient information to be submitted as a preliminary case report. The Section 106 process must be completed prior to the issuance of the final EIS.

36 CFR Part 1204 sets forth the National Register Criteria. The preceding discussion has included much of the activity required under Section 106 of the National Historic Preservation Act (NHPA). Section 106 of the National Historic Preservation Act of 1966, as amended, directs federal agencies to assess the effect of their projects on any district, site, structure, or object included in or eligible for the National Register of Historic Places. Federal agencies must obtain the review and comment of the Advisory Council on Historic Preservation before approving projects that affect such properties. As stated in 36 CFR 800.9, a project or undertaking shall be considered to have an effect:

> ". . . whenever any condition of the undertaking causes or may cause any change, beneficial or adverse, in the quality of the historical, architectural, archaeological, or cultural characteristics that qualify that property to meet the criteria of the National Register. An effect occurs when an undertaking changes the integrity of location, design, setting, materials, workmanship, feeling, or association of the property that contributes to its significance in accordance with the National Register criteria. An effect may be direct or indirect. Direct effects are caused by the undertaking and occur at the same time and place. Indirect effects include those caused by the undertaking that are later in time or farther removed in distance, but are still reasonably foreseeable. Such effects may include changes in the pattern of land use, population density, or growth rate that may affect properties of historical, architectural, archaeological, or cultural significance."

The procedures for meeting the Section 106 requirements are defined in the Advisory Council on Historic Preservation Regulations, "Protection of Historic Properties" 36 CFR Part 800. Along with the agency, the State Historic Preservation Officer (SHPO), and the Advisory Council on Historic Preservation, other participants in this process may include interested persons such as local governments, Indian tribes, and the public. The State Historic Preservation Officer (SHPO) is responsible for providing an inventory of cultural resources and for implementing programs to protect listed or eligible properties.

Under the requirements of the NHPA, no federal agency may issue a permit until the Section 106 process has been completed. Generally, the Section 106 review should run concurrently with the NEPA review process and be initiated early in the facilities planning process. Most of the Section 106 process has already been described earlier in the sections dealing with the National Register and with the determination of adverse effects.

The Advisory Council on Historic Preservation has a key role in the preservation of historic and cultural resources. Its activities have been spelled out previously. It has been the experience of the author that the Council takes its role seriously and has made a substantial contribution to historic preservation in the United States.

President Jimmy Carter, in his Memorandum on Environmental Quality and Water Resources Management dated July 12, 1978, directed the Advisory Council to promulgate regulations for implementing the National Historic Preservation Act. He further directed federal agencies with consultative responsibilities under the Act to publish separate procedures for implementing the Advisory Council's regulations.

Section 800.11 of the Advisory Council regulations 36 CFR 800, developed in response to the Presidential Memorandum, provides that certain responsibilities of individual federal agencies may be met by counterpart regulations jointly drafted by that agency and the executive director of the Advisory Council and approved by the chairman of the Advisory Council.

36 CFR 800 further provides that any federal agency finding that any proposed activities may have adverse effects on a property or on a property determined to be eligible for the National Register must allow the Advisory Council opportunity to comment on the undertaking. Advisory Council comments and recommendations are not binding. On October 1, 1986, the Council published revisions to these regulations implementing Section 106 of the National Historic Preservation Act. This final rule-making established the Council's revised regulations governing the process of review and comment upon federally supported undertakings that affect historic properties.

Executive Order 11593 (Protection and Enhancement of Cultural Environments) calls upon federal agencies to comply with NHPA and with the Advisory Council recommendations for implementing them. Several agencies have developed such guidelines. For example, the Corps of Engineers issued regulations (33 CFR Part 235, 1981) for implementing them. These regulations establish the procedures to be followed by the U.S. Army Corps of Engineers in its regulatory program in order to comply with the National Historic Preservation Act, implementing regulations, and Executive Orders for the protection of historic and cultural properties. The regulations were jointly drafted with the Advisory Council on Historic Preservation as counterpart regulations pursuant to 36 CFR 800.11. The regulations make it the district engineers' responsibility to comply with them on all permit requests in their areas.

A number of other federal laws and regulations that affect cultural and historic resources are shown below. Most of them are intended to supplement or enforce the ones previously cited. Some are as follows:

- Executive Order 11593, Protection and Enhancement of the Cultural Environment, 16 U.S.C. 470 (Sept. 1, 1971)—indirectly referred to earlier in this section requires federal agencies to survey and nominate sites on their properties to the National Register; also requires checks on any actions funded, licensed, or executed by the federal government to determine eligibility of any of the properties by the *Federal Register*.
- Archaeological and Historic Preservation Act of 1974, PL 93-291, 16 U.S.C. 469 et seq.—Act emphasizes the recovery and preservation of historic and archaeological data before being lost as a result of federal activities, licenses, or permits.
- 36 CFR Part 1204—National Register Criteria—describes criteria for inclusion in the National Register.
- Antiquity Act of 1906, PL 59-209, 34 Stat. 225; 16 U.S.C. 431–433—earliest Act of its type; protects historic and prehistoric ruins or monuments on federal lands.
- The Reservoir Salvage Act of 1969, PL 86-523, 74 STA 7.220; 16 U.S.C. 469–469c—provides for archaeological surveys of land that would be flooded by construction of a dam.

In addition, appropriate state and local regulations concerning historic and cultural resources must be satisfied where applicable. The extent and direction of these regulations will vary between states and localities. States generally cover archaeological finds, while local ordinances relate to historic and cultural sites. Each state has a State Historic Preservation Officer (SHPO) who should be consulted for existing information on any project site, as well as state requirements for surveys.

12.2   EIS CONTENTS

The state of Maryland Department of Natural Resources (1974) lists the following requirements for the state of Maryland equivalent of an EIS:

"Historic Considerations

1. Inventory historical features within or adjacent to the site including:
   a. cemeteries
   b. buildings
   c. parks
   d. bridges, canals, etc
   e. battlefields
   f. past land uses

Archaeological Considerations

1. Inventory archaeological features within or adjacent to the site including:
   a. village sites
   b. trails
   c. artifact sites
   d. burial grounds
   e. battlefields
   f. other considerations"

The Maryland Historical Trust (1981) has divided archaeological work into the five following categories:

1. Background research or prefield work preparation.
2. Preliminary archaeological reconnaissance.
3. Intensive archaeological survey.
4. Preliminary site examination.
5. Full scale excavation.

An EIS usually includes the first item. In addition, some of the other items may be included, depending upon the specific situation. A brief discussion of activities included in first two items follows (MD Historical Trust, 1981)

12.2.1   BACKGROUND RESEARCH

For both historic and prehistoric archaeology, inventory known sites to develop predictions of the locations of historic and prehistoric archaeological sites through a search of relevant documents and maps prior to the initiation of the fieldwork. The purposes are as following:

- Identification of potential sites.
- Description of historic and prehistoric settlement patterns and land-use trends.
- Identification of possible areas of racial and ethnic diversity.
- Identification of industry, commerce, and growth in the study area and its relationship to regional patterns.
- A predictive model for prehistoric site location based on available water, soil drainage, lithic resources, topography, cultural processes, and known site locations.

For urban locations, the following goals are suggested:

- Determination of evolutionary growth of the area.
- Identification of the range of social and economic activities.
- Identification of social groups.
- Identification of the types of property.
- Identification of past construction activities which might have destroyed various types of archaeological resources.
- Determination of significant types of historic archaeological sites in the project area.

12.2.2   PRELIMINARY ARCHAEOLOGICAL RECONNAISSANCE

The purpose of this reconnaissance is to locate and describe significant or potentially significant sites and areas by conducting an on-the-ground surface and sub-surface examination of the study area adequate to assess the nature and number of archaeological resources present. The reconnaissance should be sufficiently thorough to indicate if any potentially significant archaeological resources are present, but not necessarily of such intensity as to locate all sites. In the case of historic sites, where information is available and relevant, socioeconomics, race and ethnicity, and their relationship to the region and standing structures should be determined. The project impact on sites (direct and indirect) should be determined along with the need for further work, as well as any possible mitigating measures. Fieldwork consists of a selective examination of the project area including the use of shovel test-pitting, usually within an explicit sampling framework.

12.3   SURVEY METHODOLOGY

The following discussion presents suggested methodologies for cultural and historic resource effects from projects. The EIS review process involves the identification of any significant cultural resources, particularly those included in or eligible for the National Register of Historic Places; the determination of how and to what extent these resources will be impacted; and whether adequate steps have been taken to protect or mitigate these resources. This process is satisfied by adhering to the appropriate state and federal regulations and procedures and through consultation with the appropriate State Historic Preservation Officer (SHPO) and the Advisory Council on Historic Preservation (ACHP).

A typical cultural resources investigation may include some or all of the following methods:

- Examine available documents to determine adequacy of collected information on known cultural resources and potential impacts.
- Consult with the SHPO to determine the adequacy of existing cultural resource information and the need for original field work (i.e., surveys).
- Conduct literature searches at the state historical societies and SHPO headquarters for existing site locations, including a check of the *National Register of Historic Places* and official state listings.
- Contact local historical societies, universities, and colleges, amateur archaeologists, knowledgeable persons, and planning officials to locate potentially significant sites.
- Map known and/or documented cultural resources sites on USGS topographic base maps.
- Conduct windshield and/or intensive on-foot surveys to locate undocumented cultural resource sites and areas of high site potential in primary impact areas, check previously documented sites for accuracy, undertake shovel and/or auger tests where ground cover precludes visibility or where there is a likelihood of buried cultural resources.
- Conduct windshield and/or intensive on-foot surveys in areas designated as secondary impact areas.
- Compile a photographic log of each located site and its setting.
- Map newly located cultural resource sites on USGS topographic base maps; prepare site-specific maps.
- Determine significance of identified cultural resources; recommend testing of archaeological sites, if necessary, to determine significance; compile *National Register of Historic Places* inventory-nomination forms.
- Determine potential primary and secondary impacts on identified cultural resources.
- Determine possible mitigative measures and/or recommendations.
- When necessary, function as liaison between the SHPO, ACHP, and the federal agency involved.
- Prepare a final report in accordance with professional guidelines for cultural resources management, including a complete set of exhibits (maps, photographs, and survey records).

These approaches are described in more detail as follows and are broken down into literature review, field work, and impact determination.

12.3.1   IDENTIFICATION OF RESOURCES—LITERATURE REVIEW

12.3.1.1   National Register Resources

Identification of cultural resources listed in the *National Register of Historic Places* is accomplished by a review of the listing published in the *Federal Register*. Such resources usually are identified for the entire area of potential primary and secondary impacts. In addition, contact is made with the SHPO to obtain information on the resources of the project area and to discuss the proposed project with him/her personally.

12.3.1.2   Resources That May Be Eligible for the National Register of Historic Places

A literature and document review is conducted to identify the cultural resources in and adjacent to the project area that may be eligible for the National Register. Information on known archaeological sites and historic or architecturally significant structures or sites in the area is obtained from written documents and from consultations with knowledgeable individuals and agencies. Early atlases, road survey records, tax records, and property maps provide useful information. Local informants, historians, and collectors also are contacted for information on archaeological, architectural, and historic sites. Consultation is made with the SHPO, knowledgeable agencies, historic and archaeological societies, members of anthropology and history departments of local universities, and other knowledgeable individuals or groups.

The natural and geologic history of the area is investigated for clues to the location of possible archaeological sites. The physiography, past and present water resources, soils, drainage, historic vegetation, and former land uses in the project area are analyzed from an archaeological perspective to locate areas that are likely to contain prehistoric remains.

12.3.2   FIELD RECONNAISSANCE

12.3.2.1   Historic and Architectural Resources

Consultation with the SHPO should provide information on historic and architectural resource surveys previously conducted in the project area. If an adequate survey has not been conducted previously, the SHPO will indicate whether it should be undertaken.

The areas of potential primary and secondary impacts of the project are investigated. During a windshield survey, the architectural or cultural historian examines buildings, sites, or districts in the project area and its environs for evidence of historic integrity, architectural merit, unusual architectural details, and preservation. Preliminary literature and atlas review and other consultations provide additional information about the historic age and significance of the structure, site, or property. From the survey, the architectural or cultural historian compiles an architectural or historic resources inventory.

Structures are included in the historic standing structures inventory according to their inferred or documented dates of construction, their architectural merit, their relevance to the general history of the area, and their local, state, or national significance. Structures determined to be of historic or architectural value are photographed, described briefly, and located on a base map. Photographs of each historic or architecturally significant structure are included in the EIS. Evaluation of the significance of each historic structure is provided, and those that may be eligible for the National Register of Historic Places are identified. A report of findings may be submitted to the SHPO for concurrence. Nomination reports are prepared for structures or districts determined in consultation with the SHPO to meet the criteria of eligibility for the National Register.

12.3.2.2 Archaeological Resources

There have been few comprehensive archaeological surveys conducted on a statewide basis. Limited information may be on file at universities and at offices of the SHPO as a result of previous surveys undertaken in or adjacent to the project area. Thus, to identify the archaeological resources that may be eligible for listing in the National Register of Historic Places, a visual archaeological survey of the project site is required. Information obtained from the preliminary literature, map review, and consultation with informants will indicate whether all or only sections of the project site require survey. Special situations that preclude the conducting of a survey may apply, or may limit its scope. These might include previous deep disturbance of the earth on all or part of the project site, or previous emplacement of fill on the site to such a depth that construction activities on the site will not extend into the undisturbed earth beneath the fill.

Based on the literature review and informant contacts, all known historic and prehistoric archaeological sites in the area of impact are identified, briefly described, and mapped. Exact locations of archaeological sites, if known, remain as in-house data. Sites are mapped at a scale appropriate for planning purposes. Areas of potential archaeological sensitivity are identified and a research design is prepared for a Step I archaeological field survey.

The following is a typical Step I archaeological field reconnaissance:

1. Prior to conducting the survey, aerial photographs and infrared aerial photographs are scrutinized for clues to archaeological sites and former land uses on the site.
2. The archaeological survey is conducted by means of traverses across the area to be surveyed at intervals designed to locate buried resources according to terrain, archaeological potential, soil conditions, and other relevant criteria. Ideally, every meter of exposed ground surface is observed.
3. Where there is surface indication of archaeological remains (historic or prehistoric), the location of the remains is noted and mapped. Controlled surface collections or sample inventories are made at sites visible on the ground surface.
4. Where the ground surface is obscured by natural vegetation or man-made cover, one or more of the following procedures may be required when

literature review and analysis of environmental history indicate the presence of archaeological remains, or a potential for the occurrence of buried cultural material:

- Shovel test pits are spaced at intervals most likely to uncover archaeological evidence. By means of such testing, an attempt is made to delineate the nature and extent of the remains.
- In areas where shovel testing cannot be conducted because of high water table, marshy conditions, or previous filling operations, auger testing may be appropriate. In cases where deep fill has been placed over a previously undisturbed ground surface and construction activities will disturb ground beneath the fill, soil borings may be needed.
- Where impermeable man-made or other ground cover prevents testing of areas with archaeological potential, such conditions are noted.

5. Where there is no documentary or superficial evidence for the occurrence of archaeological resources and the archaeological potential is not great:
- Shovel tests are made at 16 to 30 meters (50 to 100 ft) intervals, depending on topography and other field conditions.
- In swampy areas, where the water table is high, and in filled areas, a series of auger borings may be necessary.

A comprehensive photographic record of the field survey is kept. Written daily field records also are maintained. Soil profiles are made of test pits, and any features uncovered are measured, mapped, and recorded. Artifacts recovered are washed, analyzed, and temporarily stored. A repository for archaeological material is determined in consultation with the appropriate SHPO.

A separate report of the field studies is prepared in accordance with guidelines for the preparation of archaeological reports promulgated by the National Park Service, Denver Service Center, Denver, CO, or the specific state involved. If necessary, historic and archaeological site survey forms are completed and copies of the forms are provided to the SHPO for inclusion in the statewide comprehensive site survey. In consultation with the SHPO, a determination is made of the sites that appear to be eligible for the National Register of Historic Places.

The final report of the archaeological survey includes:

- A clear, concise text including a description of the project area; a brief description of the geologic history, physiography, and vegetation of the area; and a short discussion of the history and prehistory of the area.
- A complete description of survey methodology, archaeological field strategy, and the techniques employed.
- An inventory of the known cultural resources at the project site and within a 0.25 mi radius of the project site; an inventory, description, and evaluation of cultural resources found by the survey in the project area.
- Indication of sections of the project site, if any, where an intensive Phase II cultural resources survey is recommended. Recommendations for conduct of such work and for avoidance of adverse effects to significant cultural remains discovered by the Stage I cultural resources survey are included.

- A base map of the project area on which significant cultural resources discovered by the archaeological survey, if any, are mapped. The extent of
  areas surveyed is depicted. Graphics of relevant soil profiles uncovered by
  the survey are also included, if appropriate.

12.3.3 Assessment of Impacts

12.3.3.1 Primary Impacts

Primary impacts on historic or architectural resources are those which result from the
construction or operation of the proposed facility and would constitute either beneficial or adverse effects to sites, properties, structures, or objects that are listed or that
are determined to be eligible for listing in the National Register of Historic Places.
Adverse effects may consist of one or more of the following (36 CFR Part 800, as
amended):

- Destruction or alteration of all or part of the property.
- Isolation from or alteration of its surrounding environment.
- Introduction of visual, audible, or atmospheric elements that are out of
  character with the property or that alter its setting.
- Transfer or sale of a federally-owned property without adequate conditions
  or restrictions regarding preservation, maintenance, or use.
- Neglect of a property resulting in its deterioration or destruction.

If it is discovered that an historic or architectural site, property, structure, or
object listed in or eligible for listing in the National Register of Historic Places will
be affected by the proposed project, the SHPO should be consulted for a determination of effect. If it is determined that there will be an unavoidable adverse effect on
the National Register resource, Section 106 procedures described in the National
Historic Preservation Act of 1966 and the Advisory Council Procedures for
Protection of Historic and Cultural Properties must take place.

Primary impacts on archaeological resources may occur wherever the ground
surface is disturbed by construction activities associated with a proposed project.
Construction impacts on archaeological remains consist of potential disturbances and
destruction of sites with consequent loss of scientific information. Recommendations
are made for avoidance or mitigation of such adverse effects on archaeological
remains that appear to be eligible for the National Register of Historic Places.

12.3.3.2 Secondary Impacts

Development of or alteration to the open space that presently may surround known
historic or architectural resources in the project area and constitutes an integral part
of the historic setting may diminish the historic or architectural integrity of such
properties. Similarly, alteration of the character of potential historic districts by the
introduction of structures, objects, or land uses incompatible with the historic setting
and buildings of the district would constitute an adverse impact on the historic or
architectural quality of the district. Should adverse impacts to historic resources

listed in or eligible for the National Register of Historic Places be unavoidable, execution of a memorandum of agreement may be required as specified in the Advisory Council Procedures for the Protection of Historic and Cultural Properties, as amended.

Adverse impacts to buried prehistoric and historic archaeological resources may result from future land development related to implementation of a federally-licensed or -permitted project. Accordingly, procedures for documentation of archaeological resources and field reconnaissance will be required, as necessary.

REFERENCES

Revised guidelines for implementation of the Maryland Environmental Policy Act, Maryland Department of Natural Resources, June 15, 1974.
Guidelines for archaeological investigations in Maryland, Tech. Rep. No. 1, Maryland Historic Trust, February 1981.

13 The Man-Made Environment: Transportation

Transportation factors such as traffic and parking always should be considered in the preparation of a NEPA statement. In some cases, the existing transportation and the impacts upon it may be major factors in the study, for example, large buildings and facilities in urban areas. In others, transportation issues may be relatively minor.

The basic points to be considered include answers to the following questions:

- Are there adequate transportation and parking modes presently in existence to bring workers to the project site?
- What will be the effects on existing transportation systems?
- Will new transportation facilities be required?

The discussion should focus on the effects on such things as public transit systems capacities and constraints, vehicular congestion, capacity of existing roads, safety considerations, adequacy of parking, and so on. A new project can impact transportation systems in several ways:

- Construction and operation may place increased and perhaps unattainable demands on existing transportation systems.
- Operation may generate a demand for new facilities (e.g., noise barriers, parking facilities, etc.).
- Construction and operation may permanently displace certain transportation routes.
- Location and operation of the project may reduce transportation accessibility and access to other facilities and services.

To evaluate these potential impacts, a number of techniques are used to quantify existing levels of service and to document impacts which might result from the construction of the proposed project. Typically, the starting point is an inventory of the situation in the proposed study area, including roads, public transportation, and parking facilities. These inventories are completed through a review of community maps and plans, capital improvements budgets, interviews with local officials, field surveys, and contacts with state, regional, and local agencies.

The inventory then is compared to accepted national or local standards for levels of service to determine existing capacity. Future expansion plans are evaluated in comparison to these standards to determine future capacity levels. After the land-use and population projections are completed for the no action and action alternative proposals, the impacts on transportation are documented in comparison to these future capacity levels.

Nearly every EIS requires some consideration of transportation. Invariably, an impact of some type is anticipated, its significance depending on the nature and magnitude of construction, the level of induced growth, the location of the project, and other site-specific factors.

In the detailed discussion, the entire range of transportation facilities that would be available to the staff and visitors of the project should be described in terms of transit modes, frequency of service, and commuter patterns for employment. The EIS author also should research street and highway improvement plans affecting the alternative sites; projected road capacity vs. design capacity; traffic counts on major arteries; and the nearest commercial air, rail, and bus services. Data generally are collected from the U.S. Department of Transportation and the similar state agencies.

Because of the importance of this factor in the selection of a site, particular attention should be given to the following items:

- *Local traffic considerations*: Maps may be drawn showing major traffic arteries near each site, with emphasis on peak hour traffic volumes. The number and adequacy of parking spaces also should be indicated.
- *Transit access*: Present and planned bus routes should be described and mapped; other public mass transit systems are examined thoroughly; possible carpools for employees are considered.
- *Pedestrian access*: The pedestrian access to each site should be examined with emphasis on convenience and safety.
- *Traffic analysis*: Projected levels of traffic resulting from the project are determined and compared to the existing capacity of the transportation systems. Other potential transportation problems relating to increased frequency, higher accident potential, and the transport of hazardous materials are also identified through this analysis.

13.1   EIS EXAMPLE

As an example of the type of material in an EIS where transportation is a major factor, the next several pages present an excerpt from the Environmental Assessment on the rehabilitation of Union Station, Washington, D.C. (BREGMAN & COMPANY, 1985). The factors considered here such as public transit, traffic flow, parking for taxis and automobiles, as well as construction and operation impacts are illustrative of the considerations that must be given to transportation.

"Transportation and Circulation

Union Station is well served by Washington, D.C.'s transportation system. Two broad diagonal streets, Massachusetts and Louisiana Avenues, converge on the site, while three more major arterials, North Capitol Street, Constitution Avenue, and Delaware Avenue, pass nearby. The freeway system is within three blocks of Union Station. The eight year old rapid transit system (Metro) has a station within Union Station with lines extending directly to Silver Spring and Shady Grove in Montgomery County, Maryland and with transfer service to Prince George's County, Maryland, and Alexandria, Arlington County, and Fairfax County in Virginia. Metro provides excellent access throughout the D.C. area.

Of primary importance to Union Station are the arterial streets which provide direct access. Massachusetts and Louisiana Avenues lead to Columbus Plaza in front of the Station. H and E Streets also provide direct access and they, along with Massachusetts Avenue, connect the site with ramps to and from the Center Leg freeway. H Street provides the primary access to the parking garage behind the Station.

Metrobus routes M2, X8, 11M, and 40, terminate in the garage, with direct transfers to Metrorail below. Additional Metrobus routes that serve Union Station are D2, D4, D6, D8, 42, 44, 91, 92 and X1.

Traffic Flow

The most recent environmental assessment of Union Station (1) analyzed the impact of the Union Station bus/parking garage and southeast ramp on traffic and transportation systems involving the arterial approaches to Union Station. It addressed the transportation and circulation impacts associated with a fully utilized garage of 1295 spaces. That assessment concluded that parking garage-generated traffic growth would have no significant impact on traffic or other aspects of transportation systems. Much of the following discussion is derived from that report, supplemented with an analysis of the additional impacts projected to result from the proposed project to rehabilitate Union Station.

Levels-of-service on local streets are generally acceptable with only one key north–south arterial, North Capitol Street, experiencing a poor level of service. Recent levels of service at major intersections in the evening peak hour, as reported in the 1982 Federal Highway Administration's environmental assessment (1), are shown in the next table.

Six levels of traffic service are possible, A through F. They are used to define the traffic flow at each intersection. Level A can be characterized as free flow, with little congestion and high maneuverability. Level B indicates little congestion although traffic may be moderately heavy. Level C implies heavy traffic without significant delays. Level D indicates unstable flow and significant delays. Level E is characterized by operation at full capacity, and motorists at signals often have to wait through several red lights. Finally, Level F implies forced flow with heavy queuing of vehicles with frequent stoppages.

Existing Traffic Levels-of-Service at Major Intersections

Intersection	Approach From	Level-of-Service
North Capitol and	North	B
Massachusetts Avenue, N.W.	South	C
	West	B
	East	A
North Capitol and	North	C
H Streets, N.W.	South	D
	West	C/D
	East	B
Third and H	North	A
Streets, N.W.	South	A
	West	A
	East	A

Source: 1982 Federal Highway Administration environmental assessment.

The 1982 environmental assessment conducted by Sverdrup and Parcel shows the projected average daily traffic volumes for Union Station area roadways in 1990. Roadway volumes were projected to remain at 1980 levels under the assumption that programmed expansion of Metrorail would continue to absorb any additional demand resulting from development near Union Station. A comparison of the existing and future levels-of-service at the major intersections in the Union Station area indicated that the garage/ramp project would not have an impact sufficient to change the levels-of-service which are in most cases reasonably good for peak hour traffic.

An analysis was made of the impact of the rehabilitated Union Station on major nearby intersections. The analysis indicated that approximately 50 percent of Union Station's retail customers would already be in the Station for other purposes, and that most of the restaurant customers (60 percent of sit-down restaurants, and 80 percent of fast-food restaurants) would be in the same category. The offices, on the other hand, would be primary purpose trip generators.

Based upon trip generation rates published by the Institute of Transportation Engineers, the retail and office activities planned for Union Station are expected to generate about 5000 additional round-trip vehicular trips per day, broken down as follows:

Use	AM Peak In One Way	AM Peak Out One Way	PM Peak In One Way	PM Peak Out One Way	Daily Volume One Way	Daily Volume Two Way
Retail	63	58	144	153	3,690	1,845
Sit-down restaurant	17	9	55	34	1,498	749
Fast-food restaurant	37	27	119	102	3,871	1,936
Offices	149	28	22	109	984	492
Total	266	122	340	398	10,043	5,022
Total rounded	270	120	340	400	10,000	5,000

The approach of these vehicles to Union Station may be expected from the following directions:

Street	Percent of Traffic
North Capitol	30
H Street (West)	60
H Street (East)	5
Massachusetts Avenue (Southeast)	5

The analysis indicated that the additional traffic volume through the most heavily impacted intersection, North Capitol and H Streets, N.W., would be as follows:

Approach	a.m.	p.m.
North	81	102
South	0	0
West	162	204
East	108	360

From this, it was concluded that, with proper signal timing and related transportation system management techniques, the added traffic would produce the following levels of service at North Capitol and H Streets, N.W. in the evening peak hour:

Approach	Service Level
North	C/D
South	D
West	D
East	C

These are considered acceptable service levels by the District of Columbia. Other less travelled traffic intersections, such as North Capitol Street and Massachusetts Avenue, N.W., and Third and H Streets, N.W., would have a lesser traffic increase. Accordingly, the proposed action would not have a significant impact upon vehicular traffic on major streets servicing the Union Station area.

Parking

The next table provides a breakdown of expected users of the 1296 space garage now under construction in the Union Station complex. It also shows the potential users if the garage capacity is extended to two bays to include 1830 or to three bays to contain 2114 spaces. The Barton-Aschman estimate of user categories for the 1296 space structure includes 120 spaces (10 percent of capacity) to accommodate 360 office employees and a total of 303 spaces (25.4 percent of capacity) for retail purposes, including both employees and shoppers. On this basis, the redeveloped Union Station, including office employees, would only encompass about 35 percent of the garage capacity. This is not an unduly large percentage and therefore would not have a significant impact on the garage operation over and above that anticipated before garage construction was resumed in 1984.

Sverdrup and Parcel's 1984 analysis of parking demand, expansion feasibility, and parking fees projected a level of retail activity at Union Station which could utilize at least 700 spaces, requiring a garage with 1640 spaces in 1986 or shortly thereafter. On the basis of overall projected growth at Union Station, including increased Amtrak patronage, the study projected the need for a 1820 space garage in 1993 and 2000 spaces by the year 2000. Growth resulting from the Union Station redevelopment project could be considered to have a potential long-term impact of significance to the extent that it represents a potentially larger user of parking space than in near term. Yet, realistically it can also be reasoned that this is a speculative possibility whose potential significance was recognized when the present garage construction was still in the planning stage. Moreover, any difficulties resulting from a parking shortage would be compensated for by the ready availability of public transportation to and from Union Station.

For these reasons, and certainly in the near term, the proposed Union Station rehabilitation project cannot be considered to have a significant impact on parking at Union Station.

Metro

Metro officials do not anticipate a need for increasing the number of buses which now serve Union Station. If necessary, they plan to emphasize use of Metrorail to minimize above-ground congestion. Regardless, this would involve only the use of additional subway cars rather than a revised schedule. Accordingly, the proposed Union Station rehabilitation will not have a significant impact on Metro operations.

Taxi and Auto Dropoffs

Some portion of potential retail customers at Union Station will be picked up or dropped off in the Columbus Plaza area in front of Union Station.

Projected Union Station Parking Structure Users

User Category	Previous Plan[1] 1296 Spaces		2-Bay Expansion[2] 1830 Spaces		3-Bay Expansion[2] 2114 Spaces	
Office	120	(10.0%)	120	(7.0%)	175	(8.8%)
Retail	303	(25.4%)	700	(40.8%)	765	(38.5%)
Amtrak	624	(52.3%)	625	(36.4%)	750	(37.7%)
Terminal employees	0	(0%)	90	(5.2%)	90	(4.5%)
Rental car	75	(6.3%)	125	(7.3%)	125	(6.3%)
Tourmobile	0	(0%)	50	(2.9%)	50	(2.5%)
Residual[3]	70	(5.9%)	5	(0.3%)	34	(1.7%)
Circulation Reserve[4]	104	(N/A)	115	(N/A)	125	(N/A)
	1296		1830		2114	

[1]Barton-Aschman Report.

[2]Base condition used for 2-bay expansion; best condition used for 3-bay expansion.

[3]Residual category used to define capacity available for purposes other than defined categories.

[4]Retail, Amtrak, and Tourmobile parking supply (8 percent).

Source: From Sverdrup and Parcel.

Projected Parking Need, Union Station Garage

Development Category	Parking Capacity Required					
	1986		1993		2000	
	Base	Best	Base	Best	Base	Best
Office	120	175	120	175	120	175
Retail	700	765	700	765	700	765
Amtrak	495	595	625	750	770	920
Terminal employees	75	75	90	90	110	110
Rental car	100	100	125	125	125	125
Tourbus	50	50	50	50	50	50
Reserve	100	115	110	125	125	140
Capacity required	1640	1875	1820	2080	2000	2285
Demand/capacity ratio, 2 bays	0.90	1.02	0.99	1.14	1.09	1.25
Demand/capacity ratio, 3 bays	0.78	0.89	0.86	0.98	0.95	1.08

Existing garage capacity = 1296

Total capacity, 2-bay garage expansion = 1830

Total capacity, 3-bay garage expansion = 2114   (Recommended plan)

Source: Sverdrup and Parcel.

The 1982 garage/ramp assessment (1) examined the question of taxi queuing and dropoffs. It determined by survey that the maximum accumulation of cabs was 30 to 35 at any one time during the 3:00 p.m. through 6:40 p.m. peak period. The longest queues lasted only 10 to 15 minutes and dissipated quickly once the passengers had been accommodated. Of particular interest was the volume of cabs arriving (40 percent) and leaving (70 percent) with no passengers. This indicated there should be ample taxi service for retail customers at Union Station with little or no change necessary in existing taxi service patterns.

Private automobile pickups and dropoffs of potential retail customers will most likely occur at an elevated rear access using the First Street ramp. This is an area well removed from the taxi queuing area on the east side of Union Station and therefore it will have no significant impact upon transportation and circulation at Union Station.

Construction Impacts

Some degree of inconvenience to Union Station users is unavoidable during the construction period. Project planning will seek to reduce these inconveniences to a practical minimum through construction phasing and, if necessary, temporary facilities.

Station services and access to platforms, public streets, and the Metro subway must be maintained throughout the Station renovation which is expected to last 16 months. This will be accomplished by keeping the existing replacement station in service while the historic structure is renovated.

Upon completion of the restoration and renovation work in the historic structure and the creation of a new passenger station within and adjoining, station operations will be relocated to the renovated Union Station. Passengers will have access to the trains and all platforms from the rear of the connecting structure.

Construction activities are not expected to affect bus, taxi, passenger car, or subway access to the existing temporary structure. No change in traffic patterns or existing parking is anticipated. Almost 100 percent of the work is to be accomplished inside the historic structure to which the public has no access.

In summary, the increase in traffic resulting from the renovation and rehabilitation of Union Station will not have a major adverse impact on traffic circulation and parking in the area of the Station. The proposed project will generate an estimated 5000 vehicular round trips a day but the increased peak-hour traffic in the most heavily-impacted intersection, North Capitol and H Street, N.W., will fall within an acceptable service level. The incremental increase in parking demand will not have a significant, unanticipated near-term impact upon parking or upon Metro, taxi, or auto dropoff activities. The construction operations required to carry out the project will not have an impact upon bus, taxi, passenger car, or subway access to the replacement station.

All other alternatives discussed in Section 2.1 would have lesser impacts on traffic circulation or parking than will the selected alternative."

REFERENCE

Environmental Assessment of Rehabilitation of Union Station as a Rail/Commercial/Office Facility, under subcontract to James W. Collins and Associates, BREGMAN & COMPANY, February 1985.

14 The Man-Made Environment: Socioeconomics

14.1 ORIGIN

In 1973, the Council on Environmental Quality issued guidelines for a NEPA process that considered the importance of socioeconomics in an EIS. The following key discussion in the CEQ guidelines brought socioeconomics into the picture:

"Secondary or indirect, as well as primary or direct, consequences for the environment should be included in the analysis. Many major Federal actions, in particular those that involve the construction or licensing of infrastructure investments, e.g., highways, airports, sewer systems, water resource projects, etc., stimulate or induce secondary effects in the form of associated investments and changed patterns of social and economic activities. Such secondary effects, through their impacts on existing community facilities and activities, through inducing new facilities and activities, or through changes in natural conditions, may often be even more substantial than the primary effects of the original action itself. For example, the effects of the proposed action on population and growth may be among the more significant secondary effects. Such population and growth impacts should be estimated if expected to be significant and an assessment made of the effect of any possible change in population patterns or growth upon the resource base, including land use, water, and public services, of the area in question."

The 1978 CEQ regulations defined the human environment as follows:

"Human Environment" shall be interpreted comprehensively to include the natural and physical environment and the relationship of people with that environment. This means that economic or social effects are not intended by themselves to require preparation of an environmental impact statement. When an environmental impact statement is prepared and economic or social and natural or physical environmental effects are interrelated, then the environmental impact statement will discuss all of these effects on the human environment."

14.2 ITEMS INCLUDED IN SOCIOECONOMICS

Socioeconomics covers a wide range of topics. Basically, they include those relating to human relationships and interactions, with the emphasis being on economic effects.

The socioeconomic environment in which project planning takes place is largely affected by induced growth and changes in existing land-use conditions. Growth within an area generates changes in population which, in turn, induce changes in the economy and how land is used. Induced growth also alters the demand for community services and the use of utilities, transportation facilities, and energy which, in turn, affect changes in the infrastructure of the community.

NEPA studies describe both the social and economic environment as it exists prior to the proposed project, as well as the impacts of the project on that social and economic environment. Where mitigating measures are feasible, they are included. Examination of the issues and identification of the effects on the social and economic systems are an integral part of the analysis.

Because socioeconomics is a very broad category, it has been divided in this discussion into the following subsections:

- Demography.
- Economic Base.
- Local government finances.
- Land use.
- Housing.
- Community services.
- Recreation.
- Aesthetics.

Each of these elements will be discussed on the following pages.

14.2.1 DEMOGRAPHY

The social makeup of the proposed alternative sites can be described through a review of demographic data. Topics to be addressed include population estimates and projections, the labor force and its employment level, rate, and nature of population changes, median age, median household income, age structure, and the proportion of households below the poverty level. Demographic data may be obtained from the U.S. Bureau of the Census, the state, and the local county planning boards.

The population estimation and forecast approach provides:

- Estimates of the existing population of the study area and the relevant area impacted by the proposed action.
- Projections of population changes induced by the proposed action and other alternatives under consideration.

In many cases, the preparation of NEPA studies does not require original population projections. Instead, one of several sets of projections prepared by various federal, state, regional, and/or local agencies are used. For these instances, existing projection methodologies and assumptions are evaluated to ascertain their reliability and accuracy. Methodologies that may be used include:

- Component and cohort survival analysis.
- Trend analysis.
- Saturation analysis.
- Ratio analysis.
- Segment analysis.

They all are rather complex and will not be discussed further.

In addition to general population projection techniques, a recurrent need for many alternatives is the distribution or disaggregation of larger area projections into small units that can be aggregated to the study area level. Some of the disaggregation techniques used in NEPA projects include:

- Aerial photo and segment analysis.
- Indirect indicators (e.g., gas and electric meter connections).

When aerial photos can be obtained for the total areas to be disaggregated, segment analysis is utilized to tally and proportion recent development in segments of the study area. Recent air photos also are extremely useful in the detailed land-use projection process and the identification of areas where development may be induced.

For indirect indicators, one may rely on gas and electric meter connections and other utility connections, building permits, and reviews of residential development when assessing the population growth that has occurred since the most recent population estimate. These techniques should be used cautiously because site plan review and the issuance of building permits are not necessarily synonymous with constructed units and population growth.

14.2.2 Local Economic Base

An economic profile of the project area and its surrounding region includes a description of the labor force, employment and unemployment characteristics, economic activity, and trends in each major economic sector. It also may include descriptions of local and regional income characteristics such as total personal income, per capita income, median or average household income, median household effective buying income, distribution of household income by income category, and selected salary and wage data by industry. The area income data are compared to regional or state income data to assess the relative position of the community residents. Primary data sources for the above information are:

- Local and state employment commissions.
- Chambers of Commerce.
- Local council of governments.
- Bureau of the Census.
- Department of Labor statistics.
- Department of Commerce.

Comparative data can be used to assess the trends for each economic category. An economic base analysis utilizing location quotients determines the extent of economic diversification, that is, the ratio of basic to nonbasic industries.

The role of basic employment in generating nonbasic employment and the dependence of unemployed or employed persons generates a family of ratios called employment multiplier effects.

Data on employment levels and employment projections are considered as part of the development of an EIS. Employment types include construction-phase employment, plant operations employment, and indirect and induced employment in other economic sectors. These three components of employment are used in the EIS as inputs to measure traffic generation from the facility, air emissions, noise levels, housing needs, area-wide economic effects, and community stability.

Direct and indirect employment attributed to the project can be calculated using location quotients developed in the economic base analysis. This can be calculated for the short-term peak construction phase as well as for the long-term operation and maintenance of a proposed facility. Total direct and indirect employment data then can be used to project total direct and indirect population impacts. The induced employment resulting from the proposed action may be distributed across major industry categories using existing employment ratios and comparisons of past industry trends.

Direct income resulting from the proposed facility can be calculated using wage and salary data. This is calculated for the short-term construction phase as well as for the long-term operation and maintenance of the project. By using income multipliers from regional and state input–output models, the indirect and induced gross income can be estimated as well as the gross disposable income that will be spent in the area economy. Disposable income projections are used to develop estimates of increased sales tax revenues. Income data also are useful in assessing impacts on existing labor force conditions and population migration trends.

A key element in this study is the effect of the proposed project on the retail sales potential. What is the present situation? Will the project result in an additional market for the retail sales industry? If so, to what extent? Will additional revenue be generated by sales taxes? How will this affect the overall economy of the area? These questions, not infrequently, are overriding ones in the determination of the impacts of a proposed project.

14.2.3   LOCAL GOVERNMENT FINANCES

The analysis of local government finances relies on information about the municipal budget, sources of revenue, categories of expenditures, and tax structure to assess the impacts of proposed actions on community costs. The overall population increase and induced economic growth directly affect the amount and level of community services provided by the municipality to the area.

Analysis focuses on four main categories of community costs:

- community-borne primary costs of a proposed public facility.
- community-borne secondary service and infrastructure expansion costs.

- individual user costs and ability to pay for the service generated by a public facility when that is the subject of an EIS.
- the fiscal capacity of the local government entity funding a proposed public facility.

Because the construction of new facilities may promote residential and nonresidential development, consideration must be given in the EIS to the problems and costs of supplying sufficient community and infrastructure services to meet the induced demand. Four methods of estimating service expansion costs that are variants of fiscal impact analysis or fiscal capacity analysis may be used:

Per Capita Multiplier Method. This is the straightforward means of assigning local costs to a proposed population change. It relies on detailed demographic information (by housing type) and the average cost per person of municipal and school district operating expenses to project an annual cost assignable to a particular population change.

Service Standard Method. This method relies on national or regional standards for different service categories of municipal and school district operating and capital expenditures. Future costs that will be required as a result of growth are calculated by service category.

Proportional Valuation Method. This average costing approach differentiates between residential and nonresidential impacts on local costs and revenues. This method assigns costs attributable to the share of the real property tax base. It has great value for EIS projects in which a new facility is proposed for large scale areas zoned for industrial and commercial development.

Case Study Method. This approach relies upon site-specific investigations to determine categories of excess or deficient public service capacity and the community's ability to pay for expanded services. Population-imposed needs are projected in terms of future services demand. This understanding of where excess or deficient capacity currently exists permits recommendation of service expansion needs and anticipated costs. The case study approach allows the individualized analysis of the specific service needs and financing capabilities of a particular study area.

Fiscal Capacity Analysis. This analysis includes evaluation of the budget, both general fund and departmental budgets in the case of a public facility. Sources of revenues and the tax structure are evaluated. Categories of expenditures and a statement of bonded indebtedness are provided. The government entity's ability to float bonds is evaluated within the context of the overall capital improvements plan for the community. Other funding sources such as state and federal grants are considered before the final assessment is made on the government's ability to pay for the proposed facility.

Industrial projects are evaluated by estimating increased tax revenues such as property taxes, sales taxes, or utility taxes which will result from the projects. Based on this analysis, it is determined whether the project will generate sufficient tax revenues to pay its own way.

14.2.4   LAND USE

Land-use determinations and forecasts are an essential part of EIS work. Proposed new facilities may have significant impacts on land use. Often, these facilities are major consumers of land and may serve to encourage growth of a similar nature. Land-use compatibility, both present and future, is another concern with activities of this type.

The close relationship between land-use development and population growth mandates that the techniques used for estimating and forecasting be consistent and based on common and realistic assumptions. Typically, the development and/or review of baseline and alternative population projections serve as the basis for land-use forecasting. As is the case with population projections, the methodology used for land-use forecasting varies significantly from project to project, depending upon local characteristics, available information, and the importance of land-use impacts as an EIS issue.

The land-use determination and forecasting approach is designed to produce three major products which correspond with the three major products of the population estimation and projection techniques. These are:

- Existing land-use determination of the study area and the relevant area impacted by the proposed actions.
- Baseline (no action) forecasts of land use in the study and impact areas
- Determinations of the land-use changes induced by the proposed action and other alternatives under consideration.

To achieve these desired results, one selects the most appropriate land-use determination and forecasting techniques for the particular study area. Methodologies that have been used to evaluate land-use changes include:

Carrying Capacity Analysis. The induced growth caused by new industries necessitates that the capacity of land-related elements, principal road systems, recreational facilities and open space, water supply, and other utility systems be evaluated. The capacity of each of these is measured against the projected use.

Land Absorption (Space Loss) Analysis. The increase in population that results from the development of new industries places a demand on open space. In EIS work, one identifies probable areas to be converted, measures the extent of the loss, and assesses the consequences of this loss to the community and region. Special emphasis is placed on the loss of recreational space, prime and unique farmlands, environmentally sensitive areas, and aesthetic resources.

Land-Use Cover Analysis. The construction of a new industrial project can cause a change in the land-use cover. In the EIS, the number of acres in each existing land-use category are identified. These numbers are compared to the amount that will exist after the completion of the proposed project. A change in land cover can cause alterations in plant and wildlife species as well as aesthetic qualities.

Site Location Analysis. The spatial or geographic consequences of the project on land use are assessed. The location of the project can have serious consequences on energy consumption, road utilization, accessibility to services and other parts of the region, and the compatibility of different land uses.

Land Management Analysis. An important consideration in the assessment of land-use changes is the capability of local government agencies to operate a land management system with sufficient regulatory controls and planning tools. Evaluations are typically made from zoning ordinances, subdivision ordinances, environmental control ordinances, and comprehensive plans. In reviewing these growth management tools, it is possible to identify areas which are likely to be overlooked or inadequately monitored by local regulatory institutions. Beyond control capabilities, the past performances of these institutions must be reviewed as indicators of possible future performances. Land-use controls can have their utility greatly reduced if not properly implemented by local officials.

Plan Compliance Analysis. All alternatives in an EIS are assessed to determine their conformity with municipal, county, regional, and state plans relevant to the project. These plans generally concern land use, open space and recreation, area-wide water quality management, water supply, and transportation systems.

Per Capita Land Consumption Analysis. Based on population projections of the study area, estimates can be made of the per capita consumption of land required for residential, commercial, industrial, open space, transportation, and utility uses. Residential acreage required to accommodate the projected populations is determined based on dwelling unit type and density factors and allocated to the appropriate land areas. The level of nonresidential development is then determined as a function of population increase on a per capita basis using existing local conditions as a basis for deriving the per capita standards.

In many cases, detailed existing and future land-use plans are readily available and require only an evaluation of the assumptions and methodologies used. Of particular concern in these instances is the review of the assumptions used in both the land-use and population forecasts. Before they can be used for baseline purposes, it is necessary to ascertain whether the projections are consistent and reflect common growth factors, trends, development patterns, and constraints.

Unlike population projection methodologies, land-use forecasting techniques are not generally scientific in nature. Rather, they rely on the subjective professional judgment and experience of the staff members who perform them. As a result, land-use planners typically use a combination of the previously discussed approaches that are sensitive to local characteristics, growth factors, existing available data, and other land-use determinants. This hybrid approach utilizes the best aspects of the many techniques available and allows one to tailor each analysis to the specific requirements of each project. However, those key elements of land-use forecasting that are necessary to meet the requirements mandated by NEPA and other federal regulations are always included. These specific impact evaluations include:

- The primary land-use impacts of facility construction.
- Modified rates of development which diverge significantly from that planned by governing bodies.
- Adverse unavoidable impacts in the form of induced undesirable land-use patterns.
- The relationship between short-term uses of the land versus long-term productivity.
- Irreversible and irretrievable commitments of land through primary or secondary development.
- The development of prime and unique farmlands.

In the NEPA study, the existing land use of the areas in the vicinity of the project sites should be described with emphasis on the sites themselves. Commercial and retail establishments, parking lots, housing, etc., in the general area are described. Land-use plans for the study area and zoning designation are documented. Land-use and zoning information generally may be obtained from local agencies.

Insofar as preservation of farmland is concerned, several categories of agricultural land are recognized by the U.S. Environmental Protection Agency (EPA) and the U.S. Department of Agriculture (USDA) as worthy of protection from conversion to nonfarm land uses. This includes prime and unique farmlands of national, statewide, or local significance in agricultural production; farmlands within or contiguous to environmentally sensitive areas; farmlands that may be used for land treatment or organic wastes; and farmlands with significant capital investments that help control soil erosion and nonpoint source water pollution. These are pinpointed in the NEPA study. The evaluation of prime farmlands was discussed in an earlier chapter in this book.

14.2.5   COMMUNITY SERVICES

The study of community services begins with an examination of the infrastructure. The infrastructure at the proposed site to serve the project and the staff members who will live nearby is described in terms of water supply systems, stormwater drainage, wastewater systems, solid waste disposal, and energy utilities. The present water supply in the vicinity of the development is described with regard to source of supply, storage capacity, demand, quality of supplies, and plans for expansion. This information may be obtained from the county utilities agency and the planning agencies.

Stormwater management is described in terms of existing facilities and future requirements. Data on facilities, capacities, and expansion plans are obtained from the planning agencies and the appropriate county and city agencies.

The existing wastewater collection and treatment facilities serving the site should be discussed. Items to be described include disposal methods, service areas, collection and treatment capacity, plans for expansion, service agreements, and the quality of discharged wastewater. The necessary data may be collected from the same groups listed in the preceding two paragraphs.

The existing solid waste disposal systems operated by local jurisdictions are described in terms of the identification of sanitary landfills or incinerators vs. projected demand, service areas, average pounds per day per person, and plans for expansion. The appropriate state agency and the EPA should be contacted to determine whether or not any hazardous waste disposal sites are located in the study area.

Studies on residential energy use and supply are evaluated through contact with the local utilities departments and the various planning agencies.

Community services at each proposed site to be examined in the NEPA study include schools, police, fire, recreation, health care, and shopping facilities. Information about schools and universities are obtained from the local school systems and the nearby universities. Police, fire, and rescue services are documented. Manpower resources, locations of stations, and standards of service are examined. Data sources include the county fire marshals, the county sheriff's departments, and the police and fire departments in the areas studied.

Shopping facilities are described according to the location of major commercial centers, the sizes of shopping areas, and the distances from the proposed developments. Shopping information is collected from the planning boards.

14.2.6 RECREATION

The construction of industrial or governmental facilities may cause increases in population and changes in land use that may have adverse effects on recreational resources. Conversely, opportunities for recreation may be created. The nature and importance of the recreational impact varies greatly with different types of projects. The approach to the evaluation of primary and secondary impacts on recreational resources generally includes the following steps:

- A literature review to identify the locations, types, and characteristics of parklands in and adjacent to the project area.
- A windshield survey to document current conditions.
- A mapping program to show the proximity of existing and proposed parklands.
- An areal estimate of parklands lost or directly affected by the construction of the proposed project.
- A comparison of the capacity, demand, and quality of experience at existing and proposed recreation facilities with and without the proposed project.

In some cases, more detailed analyses of recreational resources including users surveys, modeling or recreation demand, and predictions of the present and future worth of recreational facilities may be required. The approach described herein, however, is sufficient for most EIS work unless recreational resources are identified as a key EIS issue.

14.2.7 AESTHETICS

The potential for off-site aesthetic impacts varies greatly among project types and regional characteristics. Typically, those landscapes with extensive current industrial use are judged less scenic than rural farmscapes or forests, and impacts from new facilities are likely to be less noticeable or adverse.

The aesthetic value of the landscape is being viewed increasingly as a bona fide resource and, therefore, is addressed in many instances as part of an EIS. This is especially true in areas where recreation is a major industry and tourism brings in large revenues. Expansive facilities such as surface coal mines have a great potential for aesthetic impacts on scenic landscapes; in some cases they can be avoided (as with visual buffer strips), while in other cases the adverse impacts are unavoidable.

To assess the primary and secondary impacts of a proposed facility on the aesthetics of the project area, the character of the existing landscape must be defined. This involves both a literature review of existing information describing the study area and a windshield survey. Literature is gathered from federal, state, local, and university libraries; special government programs; state and local agencies; chambers of commerce; tourist bureaus; and interest groups.

14.2.8 HOUSING

Housing in the vicinity of each alternative site is discussed in an EIS with respect to inventory, growth, low and moderate income needs, and needs for persons who work at the project. The analysis includes a housing count, housing values, and vacancy rates. Low and moderate income needs may be assessed through a review of applicable local housing assistance plans. The hotel and motel situation for visitors also should be assessed. Housing data should be available from the U.S. Bureau of the Census and the local planning agencies.

14.2.9 OTHERS

The list of factors that may be included in the socioeconomic section of an EIS is lengthy and is a direct function of both the nature of the projects and of the neighborhood. Health and social services, availability of utilities, religious buildings, solid and hazardous materials management facilities, availability of medical facilities—any or all of these may prove to be critical elements.

14.3 IMPACTS

New construction can impact community facilities and the utility and communication networks of a study area in several ways:

- Construction activities may temporarily disrupt the provision of service and/or the operation of facilities.
- Construction and operation may place increased demands on existing community facilities and services or utilities.

- Operation may generate a demand for new services or facilities (i.e., chemical fire-fighting capabilities).
- Construction and operation may permanently displace certain facilities or utilities.
- Location and operation of the project may reduce access to other facilities and services.

To evaluate these potential impacts, a number of techniques are used to quantify existing levels of service and to document impacts which might result from the construction of the proposed project. The starting point is an inventory of the facilities and services in the proposed study area, including such items as police and fire protection, educational institutions, health facilities, government administrative facilities, libraries, public utilities, and solid waste disposal and water supply facilities. These inventories are completed through the review of community facility plans and capital improvements budgets, interviews with local officials and facility operators, field surveys, and contact with state, regional and local agencies.

The inventory is then compared to accepted national or local standards for levels of service to determine existing capacity. Future expansion plans are then evaluated in comparison to those standards to determine future capacity levels. After the land-use and population projections are completed for the no action and action alternative proposals, the impacts on community facilities and services are documented in comparison to these future capacity levels.

Impacts on public services such as the educational system, health care, police and fire protection, utilities, and solid waste management are difficult to measure quantitatively. When a project induces population or demographic developments in an area, there will be an increased demand for public services. The degree to which these services are affected is closely tied to the existing capacity of the service and requires a detailed demographic analysis of the added population that will affect each service, for example, number and age of children, number of additional homes, and so on.

Educational systems may be subject to a variety of impacts. The EIS generally considers only public elementary and secondary schools. Private school effects are difficult to estimate. Local public two and four year colleges are considered, but private universities are not.

Data and predicted impacts of the new development are best obtained from the local school district and the state office of education. Baseline conditions should include not only the existing situation, but also the projected growth of the educational system without the extra load to be imposed by project-related population increases.

Health services usually have as their major impacts the crowding of existing facilities and the increased demand for health care personnel. This applies primarily to local hospitals, but also takes in other types of services such as doctors, clinics, nursing homes, and so on. Information on baseline services is best obtained from the services themselves or from the local chambers of commerce. Often, changes in the ratios of potential patients to doctors and to hospital beds are used as measures of impacts.

Similar situations prevail for police and fire protection, utilities, and solid waste management. The EIS examines the degree of availability of those services before the project occurs and then estimates the changed requirements based on the increased population owing to the project as well as the location of that population. The ability of the present services to handle the additional load is determined as well as the extent to which new services will have to be provided and their costs. The term utilities here takes in natural gas, electricity, drinking water, sanitary, and storm sewers and similar activities.

14.4 MITIGATION MEASURES

In contrast to the natural sciences, mitigation measures for socioeconomic effects vary greatly and sometimes cannot be accomplished. An examination of some of the possible approaches to mitigation, as presented below, is illustrative of that.

Demography may change as a result of a project but cannot and should not be mitigated. The economic base of an area may change, more often than not, for the better. Satellite commercial establishments may arise to take advantage of a broadened customer base. No mitigating measures are foreseen.

Local government finances may be impacted positively or negatively by a new project, its taxes, and its requirements for services. When the impact is negative, then the mitigating measures may consist of additional spending by the local government for the required additional services. This, in turn, may lead to increased taxes or fees of one type or another. Conversely, a larger tax base owing to the project may lead to a lowering of taxes and/or fees.

14.4.1 LAND USE

Land use generally follows established community master plans, although changes may be made in the master plan because of the size and scope of the project. No mitigating measures are seen here.

14.4.2 COMMUNITY SERVICES

Mitigating measures consist of the expansion of the appropriate community services to fill the added requirements imposed by the project and the new population associated with it.

Recreation impacts can be mitigated by the provision of additional recreational facilities, either by the community or by the project developer.

Aesthetics impacts can be corrected by both natural and artificial means. Often the factors causing negative impacts can be redesigned so as to eliminate the impacts.

Housing represents a situation where either the community, state, or the project developer must assist the provision of additional housing in some manner. A shortage of housing is difficult to mitigate, but often corrects itself as real estate developers answer the need.

15 Environmental Justice

15.1 PRESENT STATUS

The subject of environmental justice in all environmental activities and, especially in regard to the NEPA, has developed almost *in toto* since the first edition of this book was published. The topic is still in a state of flux and that is why it is discussed as the last chapter in this book. However, environmental justice must be considered in all present and future NEPA studies. A failure to consider this topic could result in an entire EIS being declared invalid as a result of litigation. That is the primary reason for its inclusion here.

What is the reason for including environmental justice in a NEPA document? The presidential memorandum that accompanied Executive Order 12898 probably summed it up best by the statement of "... disproportionately high and adverse human health or environmental effects ... on minority populations and low-income populations in the United States and its territories and possessions. ..."

15.2 CIVIL RIGHTS ACT OF 1964

The origin of the need for including a study of environmental justice in environmental documentation goes back to Title VI of the Civil Rights Act of 1964, as amended. Title VI states, "No person in the United States shall, on the grounds of race, color, or national origin, be excluded from participation in, be denied the benefits of, or be subjected to discrimination under any program or activity receiving Federal financial assistance."

Environmental justice first gained public attention in 1982 with a landmark case involving a small, low-income, minority community in Warren County, NC. This predominantly African-American community was selected as a polychlorinated biphenyl (PCB) landfill disposal site. Despite protests and an NAACP request for an injunction based on racial discrimination, the landfill opened as planned.

As a direct result of this event, the General Accounting Office (GAO) was asked to examine hazardous waste landfill sitings in eight southern states to determine the extent to which minority communities were disproportionately exposed to environmental contamination. The GAO findings concluded that three out of every four landfills in eight states were in predominately African-American neighborhoods. The link between race and hazardous waste landfill sitings was further confirmed by a 1987 study conducted by the United Church of Christ's (UCC) Commission on Racial Justice. According to the UCC report, three out of every five African-Americans and Hispanic Americans lived in communities with uncontrolled toxic waste sites. The study determined race as the strongest deciding factor in the siting of hazardous waste facilities.

During this time period, the U.S. Environmental Protection Agency (EPA) began to receive allegations of discrimination with regards to access to public water and sewerage systems as well as to employment practices. These complaints were filed as a result of the issuance of pollution control permits by state and local governments that received EPA funding.

15.3   EXECUTIVE ORDER 12898

The next major development in environmental justice came on February 11, 1994 when President Clinton issued Executive Order 12898, "Federal actions to address environmental justice in minority populations and low-income regulations." This executive order was accompanied by a memorandum from the president to the heads of all departments and agencies that explained and emphasized the executive order.

The memorandum stated that the "order is designed to focus Federal attention on the environmental and human health conditions in minority communities and low-income communities with the goal of achieving Environmental Justice. That order is also intended to promote nondiscrimination in Federal programs substantially affecting human health and the environment, and to provide minority communities and low-income communities access to public information on, and an opportunity for public participation in, matters relating to human health or the environment."

All federal department and agency heads were to take steps immediately as follows:

"Ensure that all programs or activities receiving Federal financial assistance that affect human health or the environment do not directly, or through contractual or other arrangements, use criteria, methods, or practices that discriminate on the basis of race, color, or national origin."

"Each Federal agency shall analyze the environmental effects, including human health, economic and social effects, of Federal actions, including effects on minority communities and low-income communities, when such analysis is required by the National Environmental Policy Act of 1969 (NEPA), 42 U.S.C. section 4321 *et seq.* Mitigation measures outlined or analyzed in an environmental assessment, environmental impact statement, or record of decision, whenever feasible, should address significant and adverse environmental effects of proposed Federal actions on minority communities and low-income communities.

Each Federal agency shall provide opportunities for community input in the NEPA process, including identifying potential effects and mitigation measures in consulation with affected communities and improving the accessibility of meetings, crucial documents, and notices.

The Environmental Protection Agency, when reviewing environmental effects of proposed action of other Federal agencies under section 309 of the Clean Air Act, 42 U.S.C. section 7609, shall ensure that the involved agency has fully analyzed environmental effects on minority communities and low-income communities, including human health, social, and economic effects.

Each Federal agency shall ensure that the public, including minority communities and low-income communities, has adequate access to public information relating to human health or environmental planning, regulations, and enforcement when required under the Freedom of Information Act, 5 U.S.C. section 552, the Sunshine Act, 5 U.S.C. section 552b, and the Emergency Planning and Community Right-to-Know Act, 42 U.S.C. section 11044."

15.4 DEPARTMENT OF THE AIR FORCE GUIDE FOR ENVIRONMENTAL JUSTICE ANALYSIS WITH THE ENVIRONMENTAL IMPACT ANALYSIS PROCESS (EIAP)

The cover letter (Pohlman, 1997) that went with the distribution of this document to Air Force installations stated that,

> "This guide is based on experiences with a half dozen Environmental Impact Statements (EIS) and other environmental planning activities. Its focus is on the determination of potentially disproportionate impacts to low-income and minority populations through a ten-step process. Its full utilization is particularly geared to EISs as well as relatively complex Environmental Assessments having potential Environmental Justice considerations.
>
> In general, the most important way to achieve Environmental Justice goals is to fully involve low-income and minority populations in the Environmental Impact Analysis Process (EIAP). The populations should be engaged from the initial step in the Guide, that of scoping, throughout the entire process."

Thus, this document became the first Department of Defense (DOD) detailed statement that brought environmental justice into the EIS process. It is being utilized for that purpose by other DOD agencies such as the Corps of Engineers (Paxton, 1998). The guide presents an environmental justice flowchart for use in an EIS that is reproduced here as Exhibit 14. More details, on each of the steps shown in Exhibit 14 are presented in the text of the guide. They may be summed up as follows:

- Scoping meetings—identify minority and low-income populations.
- Impacts—determine if the proposed action causes impacts on minority or low-income populations.
- Effect of impacts—determine if the impacts are adverse.
- Geographic location of impacts—geographically map the impacts.
- People locations of impacts—identify the community of comparison (COC) and census tracts that are impacted.
- Number of affected people—determine the percent of minority and low-income population in the COC and in each census tract; decide which is greater.
- No disproportionate effect—if census tract percentage is less than COC percentage, presume no disproportionate effect and say so in a NEPA document.

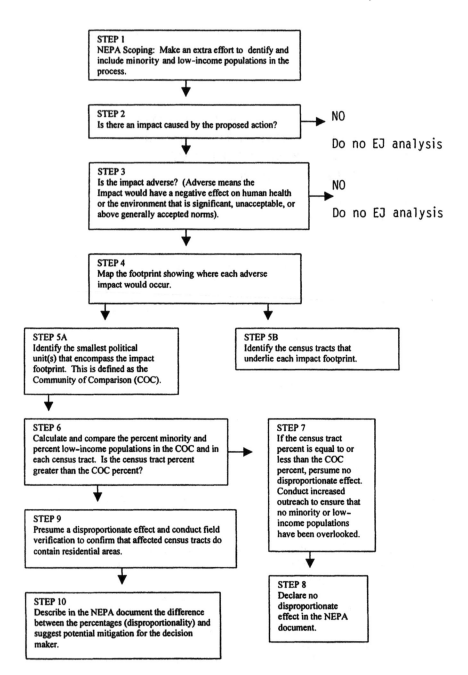

Exhibit 14   Environmental justice flowchart.

- Disproportionate effect—if census tract is greater than COC, presume that a disproportionate effect exists. Conduct field verification of residential areas in the census tract. Discuss problem in a NEPA document and suggest mitigating measures.

15.5   EPA INTERIM GUIDANCE FOR INVESTIGATING TITLE VI ADMINISTRATIVE COMPLAINTS CHALLENGING PERMITS

The U.S. Environmental Protection Agency (EPA, 1998) has issued interim guidelines on how to handle environmental justice complaints. This interim guidance was not published in the *Federal Register,* but comments were solicited with a cutoff date of May 6, 1998. The EPA had tentative plans to issue its final guidance by the late summer of 1998 but delayed the publication date.

The reason for issuing the interim guidance was that as of the date of the issuance (February 4, 1998) the Office of Civil Rights at the EPA had received 48 citizen environmental justice complaints (NAM, April 1998). Thus, the requirement for relatively rapid action on the EPA's part developed. The introduction to that guidance states that, "In the past, the Title VI complaints filed with EPA typically alleged discrimination in access to public water and sewerage systems or in employment practices. This interim guidance is intended to update the Agency's procedural and policy framework to accommodate the increasing number of Title VI complaints that allege discrimination in the environmental permitting context." The EPA background material for this guidance points out that, "The Supreme Court has ruled that Title VI authorizes EPA to adopt implementing regulations that prohibit discriminatory effects. Frequently, discrimination results from policies and practices that are neutral on their face, but have the effect of discriminating, (Department of Justice (1994)." The EPA then goes on to point out that, "as a condition of receiving EPA funds under EPA's continuing environmental program grants recipient agencies must comply with EPA's Title VI regulations." A recipient is defined as any state or its political subdivision, any public or private agency, institution, organization, entity, or any person that receives federal financial assistance either directly or indirectly. The document further states that if any part of a state or local government receives EPA funds, then all programs (including those not funded by the EPA) are subject to Title VI, for example, solid waste program funds under RCRA. If the EPA finds that discrimination exists in a recipient's permitting program, the EPA may terminate such funding.

The EPA guidelines then set up a detailed framework for processing complaints including the following:

1. Acceptance of the complaint.
2. Investigation/disparate impact assessment.
3. Rebuttal/mitigation.
4. Justification.
5. Preliminary finding of noncompliance.

6. Formal determination of noncompliance.
7. Voluntary compliance.
8. Informal resolution.

Complaints must be filed within 180 days of the alleged discriminatory act. The EPA will then accept or reject them within 20 days. Permit modifications that reduce adverse impacts are encouraged unless they result in a net increase of pollution impacts.

The investigation phase includes detailed public participation. Mitigation of the impacts are stressed, but may be waved in certain cases.

The document concludes with the following caveat:

"The statements in this document are intended solely as guidance. This document is not intended, nor can it be relied upon, to create any rights enforceable by any party in litigation with the United States. EPA may decide to follow the guidance provided in this document, or to act at variance with the guidance, based on its analysis of the specific facts presented. This guidance may be revised without notice to reflect changes in EPA's approach to implementing the Small Business Regulatory Enforcement Fairness Act or the Regulatory Flexibility Act, or to clarify and update text."

15.6   OBJECTIONS TO THE EPA INTERIM GUIDANCE

Despite, and in some cases because of, the above caveat, the EPA interim guidance has received severe criticism from a wide variety of sources. That criticism may be summed up as follows:

States—Several states have issued substantial disagreements with the EPA guidelines. They cite the following major problems:

1. The states were not consulted during the drafting of the guidelines. Their input might have avoided federal–state arguments and developed guidelines that the states could meet.
2. They would prefer guidelines to meet the policy objectives that would allow them greater flexibility to adjust programs for local needs.
3. They feel the present interim guidelines allow too much time for issues to be raised after completion of the state permitting process. They fear that this would delay many projects far too long.

These objections have been raised through the following groups:

- State attorney generals from 14 states.
- The Environmental Council of the States.

On August 10, 1998, the Environmental Council of the States (ECOS) issued their own draft alternative to the EPA's environmental justice policy. This draft, to be reviewed a month later, contained the following principles:

Principle 1—Environmental justice programs should ensure that relevant local, state, and federal agencies participate fully in the decision process and such programs must be rooted in statute or rule.

Principle 2—Environmental decisions must include meaningful public participation at the earliest possible point in time, and a meaningful effort must be made to facilitate such participation.

Principle 3—Environmental department decisions, while always subject to review in court, must be administratively final.

Principle 4—We believe that economic development and environmental protection are not mutually exclusive, but no environmental program or policy should have the effect of placing a premium on the development of "green fields" or penalizing the redevelopment of "brownfields."

Principle 5—Environmental justice programs, at whatever level of government, must use defined terms with established criteria and identified thresholds.

Principle 6—Environmental justice programs must not only treat people fairly, such programs must document that people have been fairly treated.

Principle 7—Environmental justice programs should distinguish between existing facilities that require permit renewals or modification and new facilities.

Principle 8—Environmental justice programs should contain public education elements which allow citizens to better understand environmental impacts, the state decision-making process, and industrial and commercial activities.

Principle 9—States that develop environmental justice programs which adhere to these principles should be presumed to be meeting environmental justice requirements and the burden of proof to show that they are not doing so should rest on the complainant.

It is almost certain that the final environmental justice requirements will incorporate most of both the EPA and the ECOS principles. It also is certain, given the snail's pace at which bureaucracy moves, that the final requirements will become effective in late 1999 or 2000.

One reason for state concern is that a number of projects are in an uncertain state because of possible litigation that may wind up before the Supreme Court. These especially involve incinerators and boilers in places such as Indiana, Pennsylvania, and Louisiana.

Municipalities—The U.S. Conference of Mayors has stated a concern that the interim guidance document creates potential roadblocks to the redevelopment of brownfields and other contaminated sites in minority neighborhoods and harms local government's ability to regulate polluting facilities. They are concerned that the guidance document will result in a shift in zoning control from local to federal agencies. They have asked that the guidelines be rewritten after input has been received from state and local governments.

Congress—The House Commerce Committee chairman is investigating the development of the guidelines. He is concerned with:

- The secrecy in which the guidelines were developed.
- Possible negative effects on urban development.

The committee has asked the EPA to provide it with the materials used to develop the guidelines. According to *Waste Policy Alert* (October 16, 1998), House and Senate lawmakers have inserted into EPA's FY99 budget language that effectively restricts the agency from investigating new equity complaints brought under the 1964 Civil Rights Act until EPA's guidance is finalized. EPA officials and the White House are trying to overcome this restriction (as of the date of the writing of this book).

Industry—The Business Network for Environmental Justice (BNEJ) consists of many large businesses with substantial input from the National Association of Manufacturors (NAM). BNEJ states that it wants the concept of environmental justice enforced, but fears that the EPA is going about it in the wrong way. Legal groups, such as The Washington Legal Foundation took part in a case bought by citizens of Chester, PA, against the Pennsylvania Department of Environmental Protection.

The Pennsylvania Department of Environmental Protection (DEP) had petitioned the Supreme Court to review a lower court ruling that Title VI allows citizens the right to bring civil suits against state agencies that discriminate on the basis of race or ethnic origin. The ruling came as part of a citizen suit challenging a 1995 RCRA permit granted by the DEP for a proposed remediation waste incinerator in Chester, PA. At the time, environmentalists charged that the facility would cause a disparate impact on the black community of Chester, which is home to more polluting facilities than the predominately white sections of the county.

The U.S. Supreme Court decided on August 17, 1998, to dismiss the case and ordered the 3rd Circuit Court of Appeals to vacate its ruling. The Supreme Court had heard the National Black Chamber of Commerce which asked them to strike down the lower court decision that provided a group of Pennsylvania citizens the right to file an environmental justice suit against the state under Title VI of the 1964 Civil Rights Act.

In the August 7, 1998 brief, the Chamber argued that the case was based on the "pernicious concept" of environmental justice and runs contrary to federal law and previous court decisions. They claimed that the minority effects were not so and were being used by opponents of the incinerator to try to stop the project.

15.7 NEPA CONSIDERATIONS

All of the previous discussions may be summed up as follows in so far as the relationship of environmental justice to NEPA is concerned:

- Environmental justice has a definite role in the NEPA process.
- That role is unclear at the time of the writing of this book and may involve many years of litigation.

- Authors of NEPA documents should include environmental justice considerations in their NEPA studies.

In view of the lack of firm guidelines on how to accomplish the latter point, it is suggested that personnel who are responsible for NEPA studies use the Department of the Air Force Guide, (discussed earlier in this chapter), with appropriate modifications for their particular situation. The section of this chapter that discussed the Air Force Guide listed several points in an EIS during which environmental justice should be examined. Exhibit 14, also presented earlier, gave details on ten different steps in the EIS process where environmental justice is considered. Those steps should be followed. In case of doubt, the NEPA preparer should do more rather than not enough of this work.

It is quite likely that sometime after the publication of this book, the CEQ will amend its NEPA regulations to include specific environmental justice requirements. Until that event occurs, the NEPA preparer is well-advised to include detailed analyses of this topic in his EIS.

What should go into an environmental assessment (EA)? It has been stated earlier in this book that EAs are to be brief and are to rely on existing data. That does not mean that environmental justice should be overlooked in EAs. Much of the information concerning that topic is readily available and should be included.

Possible mitigating measures should be obvious. They generally will consist of one of the following:

- Do not build the project.
- Scale it down.
- Move it elsewhere.
- Decide that project benefits will outweigh negative impacts.
- Minimize pollutants from the project to a much greater extent than standards require.

REFERENCES

The use of the disparate impact statement in administrative regulations under Title VI of the Civil Rights Act of 1964, Department of Justice, Attorney General's Memorandum for Heads of Departments and Agencies That Provide Federal Assistance, Washington, D.C., July 14, 1994.

Mullin, M., Environmental justice update, National Association of Manufacturors (NAM), Washington, D.C., April 6, 1998.

Paxton, J. E., Verbal communication, Fort Worth District, U.S. Army Corps of Engineers, 1998.

Pohlman, T. R., "Interim guidelines for environmental justice analysis with the environmental impact analysis process (EIAP), HQ USAF/ILEV, Department of the Air Force, Washington, D.C., 1997.

Interim guidance for investigating Title VI administrative complaints challenging permits, U.S. Environmental Protection Agency, Office of Civil Rights, Washington, D.C., February 4, 1998.

EPA funding bill restricts environmental equity investigations, *Waste Policy Alert*, October 16, 1998.

A The National Environmental Policy Act of 1969 as Amended

The National Environmental Policy Act of 1969, as amended:

> (Pub. L. 91-190, 42 U.S.C. 4321–4347, January 1, 1970, as amended by Pub. L. 94-52, July 3, 1975, Pub. L. 94-83, August 9, 1975, and Pub. L. 97-258, § 4(b), Sept. 13, 1982)

An Act to establish a national policy for the environment, to provide for the establishment on a Council on Environmental Quality, and for other purposes.

Be it enacted by the Senate and House of Representatives of the United States of America in Congress assembled, that this Act may be cited as the "National Environmental Policy Act of 1969."

A.1 PURPOSE

A.1.1 SECTION 2 [42 USC § 4321]

The purposes of this Act are: to declare a national policy which will encourage productive and enjoyable harmony between man and his environment; to promote efforts which will prevent or eliminate damage to the environment and biosphere and stimulate the health and welfare of man; to enrich the understanding of the ecological systems and natural resources important to the nation; and to establish a Council on Environmental Quality.

A.2 TITLE I: CONGRESSIONAL DECLARATION OF NATIONAL ENVIRONMENTAL POLICY

A.2.1 SECTION 101 [42 USC § 4331]

a. The Congress, recognizing the profound impact of man's activity on the interrelations of all components of the natural environment, particularly the profound influences of population growth, high-density urbanization, industrial expansion, resource exploitation, and new and expanding technological advances, and recognizing further the critical importance of restoring and maintaining environmental quality to the overall welfare and development of man, declares that it is the continuing policy of the federal government, in cooperation with state and local governments, and other concerned public and private organizations, to use all practicable means and measures, including financial and technical assistance, in a manner calculated to foster and promote the general welfare, to create and maintain conditions under which man and nature can exist in productive harmony, and fulfill the social, economic, and other requirements of present and future generations of Americans.

b. In order to carry out the policy set forth in this Act, it is the continuing responsibility of the federal government to use all practicable means, consist with other essential considerations of national policy, to improve and coordinate federal plans, functions, programs, and resources to the end that the nation may:

1. Fulfill the responsibilities of each generation as trustee of the environment for succeeding generations.
2. Assure for all Americans safe, healthful, productive, and aesthetically and culturally pleasing surroundings.
3. Attain the widest range of beneficial uses of the environment without degradation, risk to health or safety, or other undesirable and unintended consequences.
4. Preserve important historic, cultural, and natural aspects of our national heritage, and maintain, wherever possible, an environment which supports diversity, and variety of individual choice.
5. Achieve a balance between population and resource use which will permit high standards of living and a wide sharing of life's amenities.
6. Enhance the quality of renewable resources and approach the maximum attainable recycling of depletable resources.

c. The Congress recognizes that each person should enjoy a healthful environment and that each person has a responsibility to contribute to the preservation and enhancement of the environment.

A.2.2   SECTION 102 [42 USC § 4332]

The Congress authorizes and directs that, to the fullest extent possible:

1. The policies, regulations, and public laws of the United States shall be interpreted and administered in accordance with the policies set forth in this Act.
2. All agencies of the federal government shall:
   A. Utilize a systematic, interdisciplinary approach which will ensure the integrated use of the natural and social sciences and the environmental design arts in planning and in decision making which may have an impact on man's environment.
   B. Identify and develop methods and procedures, in consultation with the Council on Environmental Quality established by Title II of this Act, which will ensure that presently unquantified environmental amenities and values may be given appropriate consideration in decision making along with economic and technical considerations.

C. Include in every recommendation or report on proposals for legislation and other major federal actions significantly affecting the quality of the human environment, a detailed statement by the responsible official on:
   i. The environmental impact of the proposed action.
   ii. Any adverse environmental effects which cannot be avoided should the proposal be implemented.
   iii. Alternatives to the proposed action.
   iv. The relationship between local short-term uses of man's environment and the maintenance and enhancement of long-term productivity.
   v. Any irreversible and irretrievable commitments of resources which would be involved in the proposed action should it be implemented.
Prior to making any detailed statement, the responsible federal official shall consult with and obtain the comments of any federal agency which has jurisdiction by law or special expertise with respect to any environmental impact involved. Copies of such statement and the comments and views of the appropriate federal, state, and local agencies, which are authorized to develop and enforce environmental standards, shall be made available to the president, the Council on Environmental Quality, and to the public as provided by Section 552 of Title 5, United States Code, and shall accompany the proposal through the existing agency review processes.

D. Any detailed statement required under subparagraph (C) after January 1, 1970, for any major federal action funded under a program of grants to states shall not be deemed to be legally insufficient solely by reason of having been prepared by a state agency or official, if:
   i. The state agency or official has statewide jurisdiction and has the responsibility for such action.
   ii. The responsible federal official furnishes guidance and participates in such preparation.
   iii. The responsible federal official independently evaluates such statement prior to its approval and adoption.
   iv. After January 1, 1976, the responsible federal official provides early notification to and solicits the views of any other state or any federal land management entity of any action or any alternative thereto which may have significant impacts upon such state or affected federal land management entity and, if there is any disagreement on such impacts, prepares a written assessment of such impacts and views for incorporation into such detailed statement.
The procedures in this subparagraph shall not relieve the federal official of his responsibilities for the scope, objectivity, and content of the entire statement or of any other responsibility under this Act; and further, this subparagraph does not affect the legal sufficiency of statements prepared by state agencies with less than statewide jurisdiction.

E. Study, develop, and describe appropriate alternatives to recommended courses of action in any proposal which involves unresolved conflicts concerning alternative uses of available resources.

F. Recognize the worldwide and long-range character of environmental problems and, where consistent with the foreign policy of the United States, lend appropriate support to initiatives, resolutions, and programs designed to maximize international cooperation in anticipating and preventing a decline in the quality of mankind's world environment.

G. Make available to states, counties, municipalities, institutions, and individuals, advice and information useful in restoring, maintaining, and enhancing the quality of the environment.

H. Initiate and utilize ecological information in the planning and development of resource-oriented projects.

I. Assist the Council on Environmental Quality established by Title II of this Act.

A.2.3   Section 103 [42 USC § 4333]

All agencies of the federal government shall review their present statutory authority, administrative regulations, and current policies and procedures for the purpose of determining whether there are any deficiencies or inconsistencies therein which prohibit full compliance with the purposes and provisions of this Act and shall propose to the president not later than July 1, 1971, such measures as may be necessary to bring their authority and policies into conformity with the intent, purposes, and procedures set forth in this Act.

A.2.4   Section 104 [42 USC § 4334]

Nothing in Sections 102 [42 USC § 4332] or 103 [42 USC § 4333] shall in any way affect the specific statutory obligations of any federal agency to:

1. Comply with criteria or standards of environmental quality.
2. Coordinate or consult with any other federal or state agency.
3. Act or refrain from acting contingent upon the recommendations or certification of any other federal or state agency.

A.2.5   Section 105 [42 USC § 4335]

The policies and goals set forth in this Act are supplementary to those set forth in existing authorizations of federal agencies.

A.3   TITLE II: COUNCIL ON ENVIRONMENTAL QUALITY

A.3.1   Section 201 [42 USC § 4341]

The president shall transmit to the Congress annually beginning July 1, 1970, an Environmental Quality Report (hereinafter referred to as the report) which shall set forth:

1. The status and condition of the major natural, man-made, or altered environmental classes of the nation, including, but not limited to, the air, the aquatic, including marine, estuarine, and fresh water, and the terrestrial environment, including, but not limited to, the forest, dryland, wetland, range, urban, suburban, and rural environment.
2. Current and foreseeable trends in the quality, management and utilization of such environments and the effects of those trends on the social, economic, and other requirements of the nation.
3. The adequacy of available natural resources for fulfilling human and economic requirements of the nation in the light of expected population pressures.
4. A review of the programs and activities (including regulatory activities) of the federal government, the state, and local governments, and nongovernmental entities or individuals with particular reference to their effect on the environment and on the conservation, development, and utilization of natural resources.
5. A program for remedying the deficiencies of existing programs and activities, together with recommendations for legislation.

A.3.2   SECTION 202 [42 USC § 4342]

There is created in the executive office of the president a Council on Environmental Quality (hereinafter referred to as the council). The council shall be composed of three members who shall be appointed by the president to serve at his pleasure, by and with the advice and consent of the Senate. The president shall designate one of the members of the council to serve as chairman. Each member shall be a person who, as a result of his training, experience, and attainments, is exceptionally well-qualified to analyze and interpret environmental trends and information of all kinds; to appraise programs and activities of the federal government in the light of the policy set forth in Title I of this Act; to be conscious of and responsive to the scientific, economic, social, aesthetic, and cultural needs and interests of the nation; and to formulate and recommend national policies to promote the improvement of the quality of the environment.

A.3.3   SECTION 203 [42 USC § 4343]

a. The council may employ such officers and employees as may be necessary to carry out its functions under this Act. In addition, the council may employ and fix the compensation of such experts and consultants as may be necessary for the carrying out of its functions under this Act, in accordance with Section 3109 of Title 5, U.S. Code (but without regard to the last sentence thereof).
b. Notwithstanding Section 1342 of Title 31, the council may accept and employ voluntary and uncompensated services in furtherance of the purposes of the council.

A.3.4   Section 204 [42 USC § 4344]

It shall be the duty and function of the council to:

1. Assist and advise the president in the preparation of the Environmental Quality Report required by Section 201 [42 USC § 4341] of this title.
2. Gather timely and authoritative information concerning the conditions and trends in the quality of the environment both current and prospective, to analyze and interpret such information for the purpose of determining whether such conditions and trends are interfering, or are likely to interfere, with the achievement of the policy set forth in Title I of this Act, and to compile and submit to the president studies relating to such conditions and trends.
3. Review and appraise the various programs and activities of the federal government in the light of the policy set forth in Title I of this Act for the purpose of determining the extent to which such programs and activities are contributing to the achievement of such policy, and to make recommendations to the president with respect thereto.
4. Develop and recommend to the president national policies to foster and promote the improvement of environmental quality to meet the conservation, social, economic, health, and other requirements and goals of the nation.
5. Conduct investigations, studies, surveys, research, and analyses relating to ecological systems and environmental quality.
6. Document and define changes in the natural environment, including the plant and animal systems, and to accumulate necessary data and other information for a continuing analysis of these changes or trends and an interpretation of their underlying causes.
7. Report at least once each year to the president on the state and condition of the environment.
8. Make and furnish such studies, reports thereon, and recommendations with respect to matters of policy and legislation as the president may request.

A.3.5   Section 205 [42 USC § 4345]

In exercising its powers, functions, and duties under this Act, the council shall:

1. Consult with the Citizens' Advisory Committee on Environmental Quality established by Executive Order No. 11472, dated May 29, 1969, and with such representatives of science, industry, agriculture, labor, conservation organizations, state and local governments, and other groups, as it deems advisable.
2. Utilize, to the fullest extent possible, the services, facilities, and information (including statistical information) of public and private agencies and organizations, and individuals, in order that duplication of effort and expense may be avoided, thus assuring that the council's activities will not

unnecessarily overlap or conflict with similar activities authorized by law and performed by established agencies.

A.3.6   SECTION 206 [42 USC § 4346]

Members of the council shall serve full time and the chairman of the council shall be compensated at the rate provided for Level II of the Executive Schedule Pay Rates [5 USC § 5313]. The other members of the council shall be compensated at the rate provided for Level IV of the Executive Schedule Pay Rates [5 USC § 5315].

A.3.7   SECTION 207 [42 USC § 4346A]

The council may accept reimbursements from any private nonprofit organization or from any department, agency, or instrumentality of the federal government, any state, or local government, for the reasonable travel expenses incurred by an officer or employee of the council in connection with his attendance at any conference, seminar, or similar meeting conducted for the benefit of the council.

A.3.8   SECTION 208 [42 USC § 4346B]

The council may make expenditures in support of its international activities, including expenditures for:

1. International travel.
2. Activities in implementation of international agreements.
3. The support of international exchange programs in the United States and in foreign countries.

A.3.9   SECTION 209 [42 USC § 4347]

There are authorized to be appropriated to carry out the provisions of this chapter not to exceed $300,000 for fiscal year 1970, $700,000 for fiscal year 1971, and $1,000,000 for each fiscal year thereafter.

The Environmental Quality Improvement Act, as amended (Pub. L. No. 91-224, Title II, April 3, 1970; Pub. L. No. 97-258, September 13,1982; and Pub. L. No. 98-581, October 30, 1984.

A.3.10   42 USC § 4372

a. There is established in the executive office of the president an office to be known as the Office of Environmental Quality (hereafter in this chapter referred to as the office). The chairman of the Council on Environmental Quality established by Public Law 91-190 shall be the director of the office. There shall be in the office a deputy director who shall be appointed by the president, by and with the advice and consent of the Senate.
b. The compensation of the deputy director shall be fixed by the president at a rate not in excess of the annual rate of compensation payable to the deputy director of the Office of Management and Budget.

c. The director is authorized to employ such officers and employees (including experts and consultants) as may be necessary to enable the office to carry out its functions; under this chapter and Public Law 91-190, except that he may employ no more than ten specialists and other experts without regard to the provisions of Title 5, governing appointments in the competitive service, and pay such specialists and experts without regard to the provisions of Chapter 51 and Subchapter III of Chapter 53 of such title relating to classification and general schedule pay rates, but no such specialist or expert shall be paid at a rate in excess of the maximum rate for GS-18 of the general schedule under Section 5332 of Title 5.

d. In carrying out his functions, the director shall assist and advise the president on policies and programs of the federal government affecting environmental quality by:

1. Providing the professional and administrative staff and support for the Council on Environmental Quality established by Public Law 91-190.

2. Assisting the federal agencies and departments in appraising the effectiveness of existing and proposed facilities, programs, policies, and activities of the federal government, and those specific major projects designated by the president that do not require individual project authorization by Congress, that affect environmental quality.

3. Reviewing the adequacy of existing systems for monitoring and predicting environmental changes in order to achieve effective coverage and efficient use of research facilities and other resources.

4. Promoting the advancement of scientific knowledge of the effects of actions and technology on the environment and encouraging the development of the means to prevent or reduce adverse effects that endanger the health and well-being of man.

5. Assisting in coordinating among the federal departments and agencies those programs and activities which affect, protect, and improve environmental quality.

6. Assisting the federal departments and agencies in the development and interrelationship of environmental quality criteria and standards established throughout the federal government.

7. Collecting, collating, analyzing, and interpreting data and information on environmental quality, ecological research, and evaluation.

e. The director is authorized to contract with public or private agencies, institutions, and organizations and with individuals without regard to Sections 3324(a) and (b) of Title 31 and Section 5 of Title 41 in carrying out his functions.

A.3.11   42 USC § 4373

Each environmental quality report required by Public Law 91-190 shall, upon transmittal to Congress, be referred to each standing committee having jurisdiction over any part of the subject matter of the report.

A.3.12   42 USC § 4374

There are hereby authorized to be appropriated for the operations of the Office of Environmental Quality and the Council on Environmental Quality not to exceed the following sums for the following fiscal years. These sums are in addition to those contained in Public Law 91-190:

- a. $2,126,000 for the fiscal year ending September 30, 1979.
- b. $3,000,000 for the fiscal years ending September 30, 1980 and September 30, 1981.
- c. $44,000 for the fiscal years ending September 30, 1982, 1983, and 1984.
- d. $480,000 for each of the fiscal years ending September 30, 1985 and 1986.

A.3.13   42 USC § 4375

- a. There is established an Office of Environmental Quality management fund (hereinafter referred to as the fund) to receive advance payments from other agencies or accounts that may be used solely to finance:
  1. Study contracts that are jointly sponsored by the office and one or more other federal agencies.
  2. Federal interagency environmental projects (including task forces) in which the office participates.
- b. Any study contract or project that is to be financed under Subsection (a) of this section may be initiated only with the approval of the director.
- c. The director shall promulgate regulations setting forth policies and procedures for the operation of the fund.

B Council on Environmental Quality Regulations on Implementing NEPA

B.1 PART 1500—PURPOSE, POLICY, AND MANDATE

Authority: NEPA, the Environmental Quality Improvement Act of 1970, as amended (42 U.S.C. 4371 *et seq.*), Section 309 of the Clean Air Act, as amended (42 U.S.C. 7609) and Executive Order 11514, March 5, 1970, as amended by Executive Order 11991, May 24, 1977); from 43 FR 55990, November 28, 1978, unless otherwise noted.

B.1.1 SECTION 1500.1 PURPOSE

a. The National Environmental Policy Act (NEPA) is our basic national charter for protection of the environment. It establishes policy, sets goals (Section 101), and provides means (Section 102) for carrying out the policy. Section 102(2) contains action-forcing provisions to make sure that federal agencies act according to the letter and spirit of the Act. The regulations that follow implement Section 102(2). Their purpose is to tell federal agencies what they must do to comply with the procedures and achieve the goals of the Act. The president, the federal agencies, and the courts share responsibility for enforcing the Act so as to achieve the substantive requirements of Section 101.

b. NEPA procedures must ensure that environmental information is available to public officials and citizens before decisions are made and before actions are taken. The information must be of high quality. Accurate scientific analysis, expert agency comments, and public scrutiny are essential to implementing NEPA. Most important, NEPA documents must concentrate on the issues that are truly significant to the action in question, rather than amassing needless detail.

c. Ultimately, of course, it is not better documents but better decisions that count. NEPA's purpose is not to generate paperwork—even excellent paperwork—but to foster excellent action. The NEPA process is intended to help public officials make decisions that are based on understanding of environmental consequences, and to take actions that protect, restore, and enhance the environment. These regulations provide the direction to achieve this purpose.

B.1.2   Section 1500.2 Policy

Federal agencies shall to the fullest extent possible:

a. Interpret and administer the policies, regulations, and public laws of the United States in accordance with the policies set forth in the Act and in these regulations.
b. Implement procedures to make the NEPA process more useful to decision makers and the public; to reduce paperwork and the accumulation of extraneous background data; and to emphasize real environmental issues and alternatives. Environmental impact statements shall be concise, clear, and to the point, and shall be supported by evidence that agencies have made the necessary environmental analyses.
c. Integrate the requirements of NEPA with other planning and environmental review procedures required by law or by agency practice so that all such procedures run concurrently rather than consecutively.
d. Encourage and facilitate public involvement in decisions which affect the quality of the human environment.
e. Use the NEPA process to identify and assess the reasonable alternatives to proposed actions that will avoid or minimize adverse effects of these actions upon the quality of the human environment.
f. Use all practicable means, consistent with the requirements of the Act and other essential considerations of national policy, to restore and enhance the quality of the human environment and avoid or minimize any possible adverse effects of their actions upon the quality of the human environment.

B.1.3   Section 1500.3 Mandate

Parts 1500 through 1508 of this title provide regulations applicable to and binding on all federal agencies for implementing the procedural provisions of the National Environmental Policy Act of 1969, as amended (Pub. L. 91-190, 42 U.S.C. 4321 *et seq.*) (NEPA or the Act) except where compliance would be inconsistent with other statutory requirements. These regulations are issued pursuant to NEPA, the Environmental Quality Improvement Act of 1970, as amended (42 U.S.C. 4371 *et seq.*), Section 309 of the Clean Air Act, as amended (42 U.S.C. 7609), and Executive Order 11514, Protection and Enhancement of Environmental Quality (March 5, 1970, as amended by Executive Order 11991, May 24, 1977). These regulations, unlike the predecessor guidelines, are not confined to Section 102(2)(C) (environmental impact statements). The regulations apply to the whole of Section 102(2). The provisions of the Act and of these regulations must be read together as a whole in order to comply with the spirit and letter of the law. It is the council's intention that judicial review of agency compliance with these regulations not occur before an agency has filed the final environmental impact statement, or has made a final finding of no significant impact (when such a finding will result in action affecting the environment), or takes action that will result in irreparable injury. Furthermore, it is the council's intention that any trivial violation of these regulations not give rise to any independent cause of action.

B.1.4   SECTION 1500.4 REDUCING PAPERWORK

Agencies shall reduce excessive paperwork by:

a. Reducing the length of environmental impact statements [Section 1502.2(c)], by means such as setting appropriate page limits [Sections 1501.7(b)(1) and 1502.7].
b. Preparing analytic rather than encyclopedic environmental impact statements [Section 1502.2(a)].
c. Discussing only briefly issues other than significant ones [Section 1502.2(b)].
d. Writing environmental impact statements in plain language (Section 1502.8).
e. Following a clear format for environmental impact statements (Section 1502.10).
f. Emphasizing the portions of the environmental impact statement that are useful to decision makers and the public (Section 1502.14 and 1502.15) and reducing emphasis on background material (Section 1502.16).
g. Using the scoping process, not only to identify significant environmental issues deserving of study, but also to deemphasize insignificant issues, narrowing the scope of the environmental impact statement process accordingly (Section 1501.7).
h. Summarizing the environmental impact statement (Section 1502.12) and circulating the summary instead of the entire environmental impact statement if the latter is unusually long (Section 1502.19).
i. Using program, policy, or plan environmental impact statements and tiering from statements of broad scope to those of narrower scope, to eliminate repetitive discussions of the same issues (Sections 1502.4 and 1502.20).
j. Incorporating by reference (Section 1502.21).
k. Integrating NEPA requirements with other environmental review and consultation requirements (Section 1502.25).
l. Requiring comments to be as specific as possible (Section 1503.3).
m. Attaching and circulating only changes to the draft environmental impact statement, rather than rewriting and circulating the entire statement when changes are minor [Section 1503.4(c)].
n. Eliminating duplication with state and local procedures, by providing for joint preparation (Section 1506.2), and with other federal procedures, by providing that an agency may adopt appropriate environmental documents prepared by another agency (Section 1506.3).
o. Combining environmental documents with other documents (Section 1506.4).
p. Using categorical exclusions to define categories of actions which do not individually or cumulatively have a significant effect on the human environment and that are therefore exempt from requirements to prepare an environmental impact statement (Section 1508.4).
q. Using a finding of no significant impact when an action not otherwise excluded will not have a significant effect on the human environment and

is therefore exempt from requirements to prepare an environmental impact statement (Section 1508.13).

[43 FR 55990, November 29, 1978; 44 FR 873, January 3, 1979].

B.1.5   SECTION 1500.5 REDUCING DELAY

Agencies shall reduce delay by:

a. Integrating the NEPA process into early planning (Section 1501.2).
b. Emphasizing interagency cooperation before the environmental impact statement is prepared, rather than submission of adversary comments on a completed document (Section 1501.6).
c. Ensuring the swift and fair resolution of lead agency disputes (Section 1501.5).
d. Using the scoping process for an early identification of what are and what are not the real issues (Section 1501.7).
e. Establishing appropriate time limits for the environmental impact statement process [Sections 1501.7(b)(2) and 1501.8].
f. Preparing environmental impact statements early in the process (Section 1502.5).
g. Integrating NEPA requirements with other environmental review and consultation requirements (Section 1502.25).
h. Eliminating duplication with state and local procedures by providing for joint preparation (Section 1506.2) and with other federal procedures by providing that an agency may adopt appropriate environmental documents prepared by another agency (Section 1506.3).
i. Combining environmental documents with other documents (Section 1506.4).
j. Using accelerated procedures for proposals for legislation (Section 1506.8).
k. Using categorical exclusions to define categories of actions that do not individually or cumulatively have a significant effect on the human environment (Section 1508.4) and that are therefore exempt from requirements to prepare an environmental impact statement.
l. Using a finding of no significant impact when an action not otherwise excluded will not have a significant effect on the human environment (Section 1508.13) and is therefore exempt from requirements to prepare an environmental impact statement.

B.1.6   SECTION 1500.6 AGENCY AUTHORITY

Each agency shall interpret the provisions of the Act as a supplement to its existing authority and as a mandate to view traditional policies and missions in the light of the Act's national environmental objectives. Agencies shall review their policies, procedures, and regulations accordingly and revise them as necessary to ensure full compliance with the purposes and provisions of the Act. The phrase "to the fullest extent possible" in Section 102 means that each agency of the federal government shall comply with that section unless existing law applicable to the agency's operations expressly prohibits or makes compliance impossible.

B.2   PART 1501—NEPA AND AGENCY PLANNING

Authority: NEPA, the Environmental Quality Improvement Act of 1970, as amended (42 U.S.C. 4371 *et seq.*), Section 309 of the Clean Air Act, as amended (42 U.S.C. 7609, and Executive Order 11514 (March 5, 1970, as amended by Executive Order 11991, May 24, 1977) from 43 FR 55992, November 29, 1978, unless otherwise noted.

B.2.1   SECTION 1501.1 PURPOSE

The purposes of this part include:

a. Integrating the NEPA process into early planning to ensure appropriate consideration of NEPA's policies and to eliminate delay.
b. Emphasizing cooperative consultation among agencies before the environmental impact statement is prepared rather than submission of adversary comments on a completed document.
c. Providing for the swift and fair resolution of lead agency disputes.
d. Identifying at an early stage the significant environmental issues deserving of study and deemphasizing insignificant issues, narrowing the scope of the environmental impact statement accordingly.
e. Providing a mechanism for putting appropriate time limits on the environmental impact statement process.

B.2.2   SECTION 1501.2 APPLY NEPA EARLY IN THE PROCESS

Agencies shall integrate the NEPA process with other planning at the earliest possible time to ensure that planning and decisions reflect environmental values, to avoid delays later in the process, and to head off potential conflicts. Each agency shall:

a. Comply with the mandate of Section 102(2)(A) to "utilize a systematic, interdisciplinary approach which will insure the integrated use of the natural and social sciences and the environmental design arts in planning and in decision making which may have an impact on man's environment," as specified by Section 1507.2.
b. Identify environmental effects and values in adequate detail so they can be compared to economic and technical analyses. Environmental documents and appropriate analyses shall be circulated and reviewed at the same time as other planning documents.
c. Study, develop, and describe appropriate alternatives to recommended courses of action in any proposal which involves unresolved conflicts

concerning alternative uses of available resources as provided by Section 102(2)(E) of the Act.

d. Provide for cases where actions are planned by private applicants or other nonfederal entities before federal involvement so that:

1. Policies or designated staff are available to advise potential applicants of studies or other information foreseeably required for later federal action.

2. The federal agency consults early with appropriate state and local agencies and Indian tribes and with interested private persons and organizations when its own involvement is reasonably foreseeable.

3. The federal agency commences its NEPA process at the earliest possible time.

B.2.3   SECTION 1501.3 WHEN TO PREPARE AN ENVIRONMENTAL ASSESSMENT

a. Agencies shall prepare an environmental assessment (Section 1508.9) when necessary under the procedures adopted by individual agencies to supplement these regulations as described in Section 1507.3. An assessment is not necessary if the agency has decided to prepare an environmental impact statement.

b. Agencies may prepare an environmental assessment on any action at any time in order to assist agency planning and decision making.

B.2.4   SECTION 1501.4 WHETHER TO PREPARE AN ENVIRONMENTAL IMPACT STATEMENT

In determining whether to prepare an environmental impact statement the federal agency shall:

a. Determine under its procedures supplementing these regulations (described in Section 1507.3) whether the proposal is one which:

1. Normally requires an environmental impact statement.

2. Normally does not require either an environmental impact statement or an environmental assessment (categorical exclusion).

b. If the proposed action is not covered by paragraph (a) of this section, prepare an environmental assessment (Section 1508.9). The agency shall involve environmental agencies, applicants, and the public, to the extent practicable, in preparing assessments required by Section 1508.9(a)(1).

c. Based on the environmental assessment, make the determination whether or not to prepare an environmental impact statement.

d. Commence the scoping process (Section 1501.7) if the agency will prepare an environmental impact statement.

e. Prepare a finding of no significant impact (Section 1508.13) if the agency determines on the basis of the environmental assessment not to prepare an environmental impact statement.

1. The agency shall make the finding of no significant impact available to the affected public as specified in Section 1506.6.
2. Under certain limited circumstances that the agency may cover in its procedures under Section 1507.3, the agency shall make the finding of no significant impact available for public review (including state and area-wide clearinghouses) for 30 days before the agency makes its final determination whether or not to prepare an environmental impact statement and before the action may begin. The circumstances are:
    i. The proposed action is or very similar to one which normally requires the preparation of an environmental impact statement under the procedures adopted by the agency pursuant to Section 1507.3.
    ii. The nature of the proposed action is one without precedent.

B.2.5   SECTION 1501.5 LEAD AGENCIES

a. A lead agency shall supervise the preparation of an environmental impact statement if more than one federal agency either:
    1. Proposes or is involved in the same action.
    2. Is involved in a group of actions directly related to each other because of their functional interdependence or geographical proximity.
b. Federal, state, or local agencies, including at least one federal agency, may act as joint lead agencies to prepare an environmental impact statement (Section 1506.2).
c. If an action falls within the provisions of paragraph (a) of this section, the potential lead agencies shall determine by letter or memorandum which agency shall be the lead agency and which shall be cooperating agencies. The agencies shall resolve the lead agency question so as not to cause delay. If there is disagreement among the agencies, the following factors (which are listed in order of descending importance) shall determine lead agency designation:
    1. Magnitude of the agency's involvement.
    2. Project approval/disapproval authority.
    3. Expertise concerning the action's environmental effects.
    4. Duration of agency's involvement.
    5. Sequence of agency's involvement.
d. Any federal agency, or any state or local agency or private person substantially affected by the absence of lead agency designation, may make a written request to the potential lead agencies that a lead agency be designated.
e. If federal agencies are unable to agree on which agency will be the lead agency or if the procedure described in paragraph (c) of this section has not resulted in a lead agency designation within 45 days, any of the agencies or persons concerned may file a request with the council asking it to determine which federal agency shall be the lead agency. A copy of the request shall be transmitted to each potential lead agency. The request shall consist of:

    1. A precise description of the nature and extent of the proposed action.
    2. A detailed statement of why each potential lead agency should or
       should not be the lead agency under the criteria specified in paragraph
       (c) of this section.
  f. A response may be filed by any potential lead agency concerned within 20
    days after a request is filed with the council. The council shall determine
    as soon as possible, but not later than 20 days after receiving the request
    and all responses to it, which federal agency shall be the lead agency and
    which other federal agencies shall be cooperating agencies.

[43 FR 55992, November 29, 1978; 44 FR 873, January 3, 1979].

B.2.6   SECTION 1501.6 COOPERATING AGENCIES

The purpose of this section is to emphasize agency cooperation early in the NEPA
process. Upon request of the lead agency, any other federal agency that has jurisdiction by law shall be a cooperating agency. In addition, any other federal agency that
has special expertise with respect to any environmental issue that should be addressed
in the statement may be a cooperating agency upon request of the lead agency. An
agency may request the lead agency to designate it as a cooperating agency.

  a. The lead agency shall:
    1. Request the participation of each cooperating agency in the NEPA
       process at the earliest possible time.
    2. Use the environmental analysis and proposals of cooperating agencies
       with jurisdiction by law or special expertise to the maximum extent
       possible, consistent with its responsibility as lead agency.
    3. Meet with a cooperating agency at the latter's request.
  b. Each cooperating agency shall:
    1. Participate in the NEPA process at the earliest possible time.
    2. Participate in the scoping process (described below in Section 1501.7).
    3. Assume, on request of the lead agency, responsibility for developing
       information and preparing environmental analyses including portions
       of the environmental impact statement with which the cooperating
       agency has special expertise.
    4. Make available staff support at the lead agency's request to enhance the
       latter's interdisciplinary capability.
    5. Normally use its own funds. The lead agency shall, to the extent avail-
       able funds permit, fund those major activities or analyses requested
       from cooperating agencies. Potential lead agencies shall include such
       funding requirements in their budget requests.
  c. A cooperating agency may, in response to a lead agency's request for
    assistance in preparing the environmental impact statement (described in
    paragraph (b)(3), (4), or (5) of this section), reply that other program
    commitments preclude any involvement or the degree of involvement
    requested in the action that is the subject of the environmental impact
    statement. A copy of this reply shall be submitted to the council.

B.2.7 SECTION 1501.7 SCOPING

There shall be an early and open process for determining the scope of issues to be addressed and for identifying the significant issues related to a proposed action. This process shall be termed scoping. As soon as practicable after its decision to prepare an environmental impact statement and before the scoping process, the lead agency shall publish a notice of intent (Section 1508.22) in the *Federal Register* except as provided in Section 1507.3(e).

a. As part of the scoping process the lead agency shall:
   1. Invite the participation of affected federal, state, and local agencies, any affected Indian tribe, the proponent of the action, and other interested persons (including those who might not be in accord with the action on environmental grounds), unless there is a limited exception under Section 1507.3(c). An agency may give notice in accordance with Section 1506.6.
   2. Determine the scope (Section 1508.25) and the significant issues to be analyzed in depth in the environmental impact statement.
   3. Identify and eliminate from detailed study the issues which are not significant or which have been covered by prior environmental review (Section 1506.3), narrowing the discussion of these issues in the statement to a brief presentation of why they will not have a significant effect on the human environment or providing a reference to their coverage elsewhere.
   4. Allocate assignments for preparation of the environmental impact statement among the lead and cooperating agencies, with the lead agency retaining responsibility for the statement.
   5. Indicate any public environmental assessments and other environmental impact statements which are being or will be prepared that are related to but are not part of the scope of the impact statement under consideration.
   6. Identify other environmental review and consultation requirements so the lead and cooperating agencies may prepare other required analyses and studies concurrently with, and integrated with, the environmental impact statement as provided in Section 1502.25.
   7. Indicate the relationship between the timing of the preparation of environmental analyses and the agency's tentative planning and decision-making schedule.
b. As part of the scoping process the lead agency may:
   1. Set page limits on environmental documents (Section 1502.7).
   2. Set time limits (Section 1501.8).
   3. Adopt procedures under Section 1507.3 to combine its environmental assessment process with its scoping process.
   4. Hold an early scoping meeting or meetings which may be integrated with any other early planning meeting the agency has. Such a scoping meeting will often be appropriate when the impacts of a particular action are confined to specific sites.

c. An agency shall revise the determinations made under paragraphs (a) and (b) of this section if substantial changes are made later in the proposed action, or if significant new circumstances or information arise which bear on the proposal or its impacts.

B.2.8   SECTION 1501.8 TIME LIMITS

Although the council has decided that prescribed universal time limits for the entire NEPA process are too inflexible, federal agencies are encouraged to set time limits appropriate to individual actions (consistent with the time intervals required by Section 1506.10). When multiple agencies are involved, the reference to agency below means lead agency.

a. The agency shall set time limits if an applicant for the proposed action requests them provided that the limits are consistent with the purposes of NEPA and other essential considerations of national policy.
b. The agency may:
   1. Consider the following factors in determining time limits:
      i. Potential for environmental harm.
      ii. Size of the proposed action.
      iii. State of the art of analytic techniques.
      iv. Degree of public need for the proposed action, including the consequences of delay.
      v. Number of persons and agencies affected.
      vi. Degree to which relevant information is known and if not known the time required for obtaining it.
      vii. Degree to which the action is controversial.
      viii. Other time limits imposed on the agency by law, regulations, or executive order.
   2. Set overall time limits or limits for each constituent part of the NEPA process, which may include:
      i. Decision on whether to prepare an environmental impact statement (if not already decided).
      ii. Determination of the scope of the environmental impact statement.
      iii. Preparation of the draft environmental impact statement.
      iv. Review of any comments on the draft environmental impact statement from the public and agencies.
      v. Preparation of the final environmental impact statement.
      vi. Review of any comments on the final environmental impact statement.
      vii. Decision on the action based in part on the environmental impact statement.
   3. Designate a person (such as the project manager or a person in the agency's office with NEPA responsibilities) to expedite the NEPA process.

c. State or local agencies or members of the public may request a federal agency to set time limits.

B.3   PART 1502—ENVIRONMENTAL IMPACT STATEMENT

Authority: NEPA, the Environmental Quality Improvement Act of 1970, as amended (42 U.S.C. 4371 *et seq.*), Section 309 of the Clean Air Act, as amended (42 U.S.C. 7609), and Executive Order 11514 (March 5, 1970, as amended by Executive Order 11991, May 24, 1977), from 43 FR 55994, November 29, 1978, unless otherwise noted.

B.3.1   SECTION 1502.1 PURPOSE

The primary purpose of an environmental impact statement is to serve as an action-forcing device to ensure that the policies and goals defined in the Act are infused into the ongoing programs and actions of the federal government. It shall provide full and fair discussion of significant environmental impacts and shall inform decision makers and the public of the reasonable alternatives which would avoid or minimize adverse impacts or enhance the quality of the human environment. Agencies shall focus on significant environmental issues and alternatives and shall reduce paperwork and the accumulation of extraneous background data. Statements shall be concise, clear, and to the point, and shall be supported by evidence that the agency has made the necessary environmental analyses. An environmental impact statement is more than a disclosure document. It shall be used by federal officials in conjunction with other relevant material to plan actions and make decisions.

B.3.2   SECTION 1502.2 IMPLEMENTATION

To achieve the purposes set forth in Section 1502.1, agencies shall prepare environmental impact statements in the following manner:

a. Environmental impact statements shall be analytic rather than encyclopedic.
b. Impacts shall be discussed in proportion to their significance. There shall be only brief discussion of other than significant issues. As in a finding of no significant impact, there should be only enough discussion to show why more study is not warranted.
c. Environmental impact statements shall be kept concise and shall be no longer than absolutely necessary to comply with NEPA and with these regulations. Length should vary first with potential environmental problems and then with project size.
d. Environmental impact statements shall state how alternatives considered in it and decisions based on it will or will not achieve the requirements of Sections 101 and 102(1) of the Act and other environmental laws and policies.
e. The range of alternatives discussed in environmental impact statements shall encompass those to be considered by the ultimate agency decision maker.

   f. Agencies shall not commit resources prejudicing selection of alternatives
      before making a final decision (Section 1506.1).
   g. Environmental impact statements shall serve as the means of assessing the
      environmental impact of proposed agency actions, rather than justifying
      decisions already made.

B.3.3   Section 1502.3 Statutory Requirements
for Statements

As required by Section 102(2)(C) of NEPA, environmental impact statements
(Section 1508.11) are to be included in every recommendation or report.

   On proposals (Section 1508.23).
   For legislation (Section 1508.17).
   Other major federal actions (Section 1508.18).
   Significantly (Section 1508.27).
   Affecting (Sections 1508.3 and 1508.8).
   The quality of the human environment (Section 1508.14).

B.3.4   Section 1502.4 Major Federal Actions
Requiring the Preparation of Environmental
Impact Statements

   a. Agencies shall make sure the proposal which is the subject of an environ-
      mental impact statement is properly defined. Agencies shall use the crite-
      ria for scope (Section 1508.25) to determine which proposal(s) shall be the
      subject of a particular statement. Proposals or parts of proposals which are
      related to each other closely enough to be, in effect, a single course of
      action shall be evaluated in a single impact statement.
   b. Environmental impact statements may be prepared and are sometimes
      required for broad federal actions such as the adoption of new agency pro-
      grams or regulations (Section 1508.18). Agencies shall prepare statements
      on broad actions so that they are relevant to policy and are timed to coin-
      cide with meaningful points in agency planning and decision making.
   c. When preparing statements on broad actions (including proposals by more
      than one agency), agencies may find it useful to evaluate the proposal(s)
      in one of the following ways:
      1. Geographically, including actions occurring in the same general loca-
         tion, such as a body of water, region, or metropolitan area.
      2. Generically, including actions which have relevant similarities, such as
         common timing, impacts, alternatives, methods of implementation,
         media, or subject matter.
      3. By stage of technological development including federal or federally
         assisted research, development, or demonstration programs for new
         technologies which, if applied, could significantly affect the quality of

the human environment. Statements shall be prepared on such programs and shall be available before the program has reached a stage of investment or commitment to implementation likely to determine subsequent development or restrict later alternatives.

d. Agencies shall as appropriate employ scoping (Section 1501.7), tiering (Section 1502.20), and other methods listed in Sections 1500.4 and 1500.5 to relate broad and narrow actions and to avoid duplication and delay.

B.3.5   SECTION 1502.5 TIMING

An agency shall commence preparation of an environmental impact statement as close as possible to the time the agency is developing or is presented with a proposal (Section 1508.23) so that preparation can be completed in time for the final statement to be included in any recommendation or report on the proposal. The statement shall be prepared early enough so that it can serve practically as an important contribution to the decision-making process and will not be used to rationalize or justify decisions already made [Sections 1500.2(c), 1501.2, and 1502.2]. For instance:

a. For projects directly undertaken by federal agencies, the environmental impact statement shall be prepared at the feasibility analysis (go–no go) stage and may be supplemented at a later stage if necessary.

b. For applications to the agency, appropriate environmental assessments or statements shall be commenced no later than immediately after the application is received. Federal agencies are encouraged to begin preparation of such assessments or statements earlier, preferably jointly with applicable state or local agencies.

c. For adjudication, the final environmental impact statement shall normally precede the final staff recommendation and that portion of the public hearing related to the impact study. In appropriate circumstances, the statement may follow preliminary hearings designed to gather information for use in the statements.

d. For informal rulemaking, the draft environmental impact statement shall normally accompany the proposed rule.

B.3.6   SECTION 1502.6 INTERDISCIPLINARY PREPARATION

Environmental impact statements shall be prepared using an interdisciplinary approach that will ensure the integrated use of the natural and social sciences and the environmental design arts [Section 102(2)(A) of the Act]. The disciplines of the preparers shall be appropriate to the scope and issues identified in the scoping process (Section 1501.7).

B.3.7   SECTION 1502.7 PAGE LIMITS

The text of the final environmental impact statements (e.g., paragraphs (d) through (g) of Section 1502.10) shall normally be less than 150 pages and for proposals of unusual scope or complexity shall normally be less than 300 pages.

B.3.8   SECTION 1502.8 WRITING

Environmental impact statements shall be written in plain language and may use appropriate graphics so that decision makers and the public can readily understand them. Agencies should employ writers of clear prose or editors to write, review, or edit statements, which will be based upon the analysis and supporting data from the natural and social sciences and the environmental design arts.

B.3.9   SECTION 1502.9 DRAFT, FINAL,
AND SUPPLEMENTAL STATEMENTS

Except for proposals for legislation as provided in Section 1506.8, environmental impact statements shall be prepared in two stages and may be supplemented.

   a. Draft environmental impact statements shall be prepared in accordance with the scope decided upon in the scoping process. The lead agency shall work with the cooperating agencies and shall obtain comments as required in Part 1503 of this appendix. The draft statement must fulfill and satisfy to the fullest extent possible the requirements established for final statements in Section 102(2)(C) of the Act. If a draft statement is so inadequate as to preclude meaningful analysis, the agency shall prepare and circulate a revised draft of the appropriate portion. The agency shall make every effort to disclose and discuss at appropriate points in the draft statement all major points of view on the environmental impacts of the alternatives including the proposed action.

   b. Final environmental impact statements shall respond to comments as required in Part 1503 of this appendix. The agency shall discuss at appropriate points in the final statement any responsible opposing view which was not adequately discussed in the draft statement and shall indicate the agency's response to the issues raised.

   c. Agencies:

     1. Shall prepare supplements to either draft or final environmental impact statements if:

       i. The agency makes substantial changes in the proposed action that are relevant to environmental concerns.

       ii. There are significant new circumstances or information relevant to environmental concerns and bearing on the proposed action or its impacts.

     2. May also prepare supplements when the agency determines that the purposes of the Act will be furthered by doing so.

     3. Shall adopt procedures for introducing a supplement into its formal administrative record, if such a record exists.

     4. Shall prepare, circulate, and file a supplement to a statement in the same fashion (exclusive of scoping) as a draft and final statement unless alternative procedures are approved by the council.

B.3.10 Section 1502.10 Recommended Format

Agencies shall use a format for environmental impact statements that will encourage good analysis and clear presentation of the alternatives including the proposed action. The following standard format for environmental impact statements should be followed unless the agency determines that there is a compelling reason to do otherwise:

a. Cover sheet.
b. Summary.
c. Table of contents.
d. Purpose of and need for action.
e. Alternatives including proposed action, [Sections 102(2)(C)(iii) and 102(2)(E) of the Act].
f. Affected environment.
g. Environmental consequences, [especially Sections 102(2)(C)(i), (ii), (iv), and (v) of the Act].
h. List of preparers.
i. List of agencies, organizations, and persons to whom copies of the statement are sent.
j. Index.
k. Appendices (if any).

If a different format is used, it shall include paragraphs (a), (b), (c), (h), (i), and (j) of this section and shall include the substance of paragraphs (d), (e), (f), (g), and (k) of this section, as further described in Sections 1502.11 through 1502.18, in any appropriate format.

B.3.11 Section 1502.11 Cover Sheet

The cover sheet shall not exceed one page. It shall include:

a. A list of the responsible agencies including the lead agency and any cooperating agencies.
b. The title of the proposed action that is the subject of the statement (and if appropriate, the titles of related cooperating agency actions), together with the state(s) and county(ies) (or other jurisdiction, if applicable) where the action is located.
c. The name, address, and telephone number of the person at the agency who can supply further information.
d. A designation of the statement as a draft, final, or draft or final supplement.
e. A one paragraph abstract of the statement.
f. The date by which comments must be received (computed in cooperation with the EPA under Section 1506.10).

The information required by this section may be entered on Standard Form 424 (in items 4, 6, 7, 10, and 18).

B.3.12   SECTION 1502.12 SUMMARY

Each environmental impact statement shall contain a summary which adequately and accurately summarizes the statement. The summary shall stress the major conclusions, areas of controversy (including issues raised by agencies and the public), and the issues to be resolved (including the choice among alternatives). The summary will normally not exceed 15 pages.

B.3.13   SECTION 1502.13 PURPOSE AND NEED

The statement shall briefly specify the underlying purpose and need to which the agency is responding in proposing the alternatives including the proposed action.

B.3.14   SECTION 1502.14 ALTERNATIVES INCLUDING THE PROPOSED ACTION

This section is the heart of the environmental impact statement. Based on the information and analysis presented in the sections on the affected environment (Section 1502.15) and the environmental consequences (Section 1502.16), it should present the environmental impacts of the proposal and the alternatives in comparative form, thus sharply defining the issues and providing a clear basis for choice among options by the decision maker and the public. In this section agencies shall:

   a. Rigorously explore and objectively evaluate all reasonable alternatives, and for alternatives which were eliminated from the detailed study, briefly discuss the reasons for their elimination.
   b. Devote substantial treatment to each alternative considered in detail including the proposed action so that reviewers may evaluate their comparative merits.
   c. Include reasonable alternatives not within the jurisdiction of the lead agency.
   d. Include the alternative of no action.
   e. Identify the agency's preferred alternative or alternatives if one or more exist in the draft statement, and identify such alternative(s) in the final statement unless another law prohibits the expression of such a preference.
   f. Include appropriate mitigation measures not already included in the proposed action or alternatives.

B.3.15   SECTION 1502.15 AFFECTED ENVIRONMENT

The environmental impact statement shall succinctly describe the environment of the area(s) to be affected or created by the alternatives under consideration. The descriptions shall be no longer than is necessary to understand the effects of the alternatives. Data and analyses in a statement shall be commensurate with the importance of the impact with the less important material summarized, consolidated, or simply referenced. Agencies shall avoid useless bulk in statements and shall concentrate effort and attention on important issues. Verbose descriptions of the affected environment are no measure of the adequacy of an environmental impact statement.

B.3.16   SECTION 1502.16 ENVIRONMENTAL CONSEQUENCES

This section forms the scientific and analytic basis for the comparisons under Section 1502.14. It shall consolidate the discussions of those elements required by Sections 102(2)(C)(i), (ii), (iv), and (v) of NEPA which are within the scope of the statement and as much of Section 102(2)(C)(iii) as is necessary to support the comparisons. The discussion will include the environmental impacts of the alternatives including the proposed action, any adverse environmental effects that cannot be avoided should the proposal be implemented, the relationship between short-term uses of man's environment and the maintenance and enhancement of long-term productivity, and any irreversible or irretrievable commitments of resources which would be involved in the proposal should it be implemented. This section should not duplicate discussions in Section 1502.14. It shall include discussions of:

a. Direct effects and their significance (Section 1508.8).
b. Indirect effects and their significance (Section 1508.8).
c. Possible conflicts between the proposed action and the objectives of federal, regional, state, and local (and in the case of a reservation, Indian tribe) land-use plans, policies, and controls for the area concerned [Section 1506.2(d)].
d. The environmental effects of alternatives including the proposed action, the comparisons under Section 1502.14 will be based on this discussion.
e. Energy requirements and conservation potential of various alternatives and mitigation measures.
f. Natural or depletable resource requirements and conservation potential of various alternatives and mitigation measures.
g. Urban quality, historic and cultural resources, and the design of the built environment, including the reuse and conservation potential of various alternatives and mitigation measures.
h. Means to mitigate adverse environmental impacts [if not fully covered under Section 1502.14(f)].

[43 FR 55994, November 29, 1978; 44 FR 873, January 3, 1979].

B.3.17   SECTION 1502.17 LIST OF PREPARERS

The environmental impact statement shall list the names, together with their qualifications (expertise, experience, professional disciplines), of the persons who were primarily responsible for preparing the environmental impact statement or significant background papers including basic components of the statement (Sections 1502.6 and 1502.8). Where possible, the persons who are responsible for a particular analysis, including analyses in background papers, shall be identified. Normally the list will not exceed two pages.

B.3.18   SECTION 1502.18 APPENDIX

If an agency prepares an appendix to an environmental impact statement, the appendix shall:

a. Consist of material prepared in connection with an environmental impact statement [as distinct from material which is not so prepared and which is incorporated by reference (Section 1502.21)].
b. Normally consist of material which substantiates any analysis fundamental to the impact statement.
c. Normally be analytic and relevant to the decision to be made.
d. Be circulated with the environmental impact statement or be readily available upon request.

B.3.19 SECTION 1502.19 CIRCULATION OF THE ENVIRONMENTAL IMPACT STATEMENT

Agencies shall circulate the entire draft and final environmental impact statements except for certain appendices as provided in Section 1502.18(d) and unchanged statements as provided in Section 1503.4(c). However, if the statement is unusually long, the agency may circulate the summary instead, except that the entire statement shall be furnished to:

a. Any federal agency which has jurisdiction by law or special expertise with respect to any environmental impact involved and any appropriate federal, state or local agency authorized to develop and enforce environmental standards.
b. The applicant, if any.
c. Any person, organization, or agency requesting the entire environmental impact statement.
d. In the case of a final environmental impact statement any person, organization, or agency which submitted substantive comments on the draft.

If the agency circulates the summary and thereafter receives a timely request for the entire statement and for additional time to comment, the time for that requestor only shall be extended by at least 15 days beyond the minimum period.

B.3.20 SECTION 1502.20 TIERING

Agencies are encouraged to tier their environmental impact statements to eliminate repetitive discussions of the same issues and to focus on the actual issues ripe for decision at each level of the environmental review (Section 1508.28). Whenever a broad environmental impact statement has been prepared (such as a program or policy statement) and a subsequent statement or environmental assessment is then prepared on an action included within the entire program or policy (such as a site-specific action) the subsequent statement or environmental assessment need only summarize the issues discussed in the broader statement and incorporate discussions from the broader statement by reference and shall concentrate on the issues specific to the subsequent action. The subsequent document shall state where the earlier document is available. Tiering may also be appropriate for different stages of actions (Section 1508.28).

B.3.21   SECTION 1502.21 INCORPORATION BY REFERENCE

Agencies shall incorporate material into an environmental impact statement by reference when the effect will be to cut down on bulk without impeding agency and public review of the action. The incorporated material shall be cited in the statement and its content briefly described. No material may be incorporated by reference unless it is reasonably available for inspection by potentially interested persons within the time allowed for comment. Material based on proprietary data which is itself not available for review and comment shall not be incorporated by reference.

B.3.22   SECTION 1502.22 INCOMPLETE
OR UNAVAILABLE INFORMATION

When an agency is evaluating reasonably foreseeable significant adverse effects on the human environment in an environmental impact statement and there is incomplete or unavailable information, the agency shall always make clear that such information is lacking.

a. If the incomplete information relevant to reasonably foreseeable significant adverse impacts is essential to a reasoned choice among alternatives and the overall costs of obtaining it are not exorbitant, the agency shall include the information in the environmental impact statement.

b. If the information relevant to reasonably foreseeable significant adverse impacts cannot be obtained because the overall costs of obtaining it are exorbitant or the means to obtain it are not known, the agency shall include within the environmental impact statement:

1. A statement that such information is incomplete or unavailable.

2. A statement of the relevance of the incomplete or unavailable information to evaluating reasonably foreseeable significant adverse impacts on the human environment.

3. A summary of existing credible scientific evidence which is relevant to evaluating the reasonably foreseeable significant adverse impacts on the human environment.

4. The agency's evaluation of such impacts based upon theoretical approaches or research methods generally accepted in the scientific community. For the purposes of this section, "reasonably foreseeable" includes impacts that have catastrophic consequences even if their probability of occurrence is low, provided that the analysis of the impacts is supported by credible scientific evidence, is not based on pure conjecture, and is within the rule of reason.

c. The amended regulation will be applicable to all environmental impact statements for which a Notice of Intent (40 CFR 1508.22) is published in the *Federal Register* on or after May 27, 1986. For environmental impact statements in progress, agencies may choose to comply with the requirements of either the original or amended regulation.

[51 FR 15625, April 25, 1986].

B.3.23   SECTION 1502.23 COST–BENEFIT ANALYSIS

If a cost–benefit analysis relevant to the choice among environmentally different alternatives is being considered for the proposed action, it shall be incorporated by reference or appended to the statement as an aid in evaluating the environmental consequences. To assess the adequacy of compliance with Section 102(2)(B) of the Act, the statement shall, when a cost–benefit analysis is prepared, discuss the relationship between that analysis and any analyses of unquantified environmental impacts, values, and amenities. For purposes of complying with the Act, the weighing of the merits and drawbacks of the various alternatives need not be displayed in a monetary cost–benefit analysis and should not be when there are important qualitative considerations. In any event, an environmental impact statement should at least indicate those considerations, including factors not related to environmental quality, which are likely to be relevant and important to a decision.

B.3.24   SECTION 1502.24 METHODOLOGY
AND SCIENTIFIC ACCURACY

Agencies shall ensure the professional integrity, including scientific integrity, of the discussions and analyses in environmental impact statements. They shall identify any methodologies used and shall make explicit reference by footnote to the scientific and other sources relied upon for conclusions in the statement. An agency may place discussion of methodology in an appendix.

B.3.25   SECTION 1502.25 ENVIRONMENTAL REVIEW
AND CONSULTATION REQUIREMENTS

     a. To the fullest extent possible, agencies shall prepare draft environmental impact statements concurrently with and integrated with environmental impact analyses and related surveys and studies required by the Fish and Wildlife Coordination Act (16 U.S.C. 661 *et seq.*), the National Historic Preservation Act of 1966 (16 U.S.C. 470 *et seq.*), the Endangered Species Act of 1973 (16 U.S.C. 1531 *et seq.*), and other environmental review laws and executive orders.

     b. The draft environmental impact statement shall list all federal permits, licenses, and other entitlements which must be obtained in implementing the proposal. If it is uncertain whether a federal permit, license, or other entitlement is necessary, the draft environmental impact statement shall so indicate.

B.4   PART 1503—COMMENTING

Authority: NEPA, the Environmental Quality Improvement Act of 1970, as amended (42 U.S.C. 4371 *et seq.*), Section 309 of the Clean Air Act, as amended (42 U.S.C. 7609), and Executive Order 11514 (March 5, 1970, as amended by Executive Order 11991, May 24, 1977), from 43 FR 55997, November 29, 1978, unless otherwise noted.

B.4.1   SECTION 1503.1 INVITING COMMENTS

a. After preparing a draft environmental impact statement and before preparing a final environmental impact statement the agency shall:
   1. Obtain the comments of any federal agency which has jurisdiction by law or special expertise with respect to any environmental impact involved or that is authorized to develop and enforce environmental standards.
   2. Request the comments of:
      i. Appropriate state and local agencies which are authorized to develop and enforce environmental standards.
      ii. Indian tribes, when the effects may be on a reservation.
      iii. Any agency that has asked to receive statements on actions of the kind proposed.
      The Office of Management and Budget Circular A-95 (revised), through its system of clearinghouses, provides a means of securing the views of state and local environmental agencies. The clearinghouses may be used, by mutual agreement of the lead agency and the clearinghouse, for securing state and local reviews of the draft environmental impact statements.
   3. Request comments from the applicant, if any.
   4. Request comments from the public, affirmatively soliciting comments from those persons or organizations who may be interested or affected.
b. An agency may request comments on a final environmental impact statement before the decision is finally made. In any case other agencies or persons may make comments before the final decision unless a different time is provided under Section 1506.10.

B.4.2   SECTION 1503.2 DUTY TO COMMENT

Federal agencies with jurisdiction by law or special expertise with respect to any environmental impact involved and agencies that are authorized to develop and enforce environmental standards shall comment on statements within their jurisdiction, expertise, or authority. Agencies shall comment within the time period specified for comment in Section 1506.10. A federal agency may reply that it has no comment. If a cooperating agency is satisfied that its views are adequately reflected in the environmental impact statement, it should reply that it has no comment.

B.4.3   SECTION 1503.3 SPECIFICITY OF COMMENTS

a. Comments on an environmental impact statement or on a proposed action shall be as specific as possible, and may address either the adequacy of the statement or the merits of the alternatives discussed or both.
b. When a commenting agency criticizes a lead agency's predictive methodology, the commenting agency should describe the alternative methodology which it prefers and why.
c. A cooperating agency shall specify in its comments whether it needs additional information to fulfill other applicable environmental reviews or con-

sultation requirements and what information it needs. In particular, it shall specify any additional information it needs to comment adequately on the draft statement's analysis of significant site-specific effects associated with the granting or approving by that cooperating agency of the necessary federal permits, licenses, or entitlements.

d. When a cooperating agency with jurisdiction by law objects to or expresses reservations about the proposal on the grounds of environmental impacts, the agency expressing the objection or reservation shall specify the mitigation measures it considers necessary to allow the agency to grant or approve applicable permit, license, or related requirements or concurrences.

B.4.4   SECTION 1503.4 RESPONSE TO COMMENTS

a. An agency preparing a final environmental impact statement shall assess and consider comments both individually and collectively, and shall respond by one or more of the means listed below, stating its response in the final statement. Possible responses are to:
   1. Modify alternatives including the proposed action.
   2. Develop and evaluate alternatives not previously given serious consideration by the agency.
   3. Supplement, improve, or modify its analyses.
   4. Make factual corrections.
   5. Explain why the comments do not warrant further agency response, citing the sources, authorities, or reasons that support the agency's position and, if appropriate, indicate those circumstances that would trigger agency reappraisal or further response.

b. All substantive comments received on the draft statement (or summaries thereof where the response has been exceptionally voluminous) should be attached to the final statement whether or not the comment is thought to merit individual discussion by the agency in the text of the text of the statement.

c. If changes in response to comments are minor and are confined to the responses described in paragraphs (a)(4) and (5) of this section, agencies may write them on errata sheets and attach them to the statement instead of rewriting the draft statement. In such cases only the comments, the responses, and the changes and not the final statement need be circulated (Section 1502.19). The entire document with a new cover sheet shall be filed as the final statement (Section 1506.9).

B.5   PART 1504—PREDECISION REFERRALS TO THE COUNCIL OF PROPOSED FEDERAL ACTIONS DETERMINED TO BE ENVIRONMENTALLY UNSATISFACTORY

Authority: NEPA, the Environmental Quality Improvement Act of 1970, as amended (42 U.S.C. 4371 *et seq.*), Section 309 of the Clean Air Act, as amended (42 U.S.C.

7609), and Executive Order 11514 (March 5, 1970, as amended by Executive Order 11991, May 24, 1977), from 43 FR 55998, November 29, 1978, unless otherwise noted.

B.5.1   SECTION 1504.1 PURPOSE

a. This part establishes procedures for referring federal interagency disagreements concerning proposed major federal actions that might cause unsatisfactory environmental effects to the council. It provides means for early resolution of such disagreements.
b. Under Section 309 of the Clean Air Act (42 U.S.C. 7609), the administrator of the Environmental Protection Agency is directed to review and comment publicly on the environmental impacts of federal activities, including actions for which environmental impact statements are prepared. If after this review, the administrator determines that the matter is "unsatisfactory from the standpoint of public health or welfare or environmental quality," Section 309 directs that the matter be referred to the council (hereafter, environmental referrals).
c. Under Section 102(2)(C) of the Act, other federal agencies may make similar reviews of environmental impact statements, including judgments on the acceptability of anticipated environmental impacts. These reviews must be made available to the president, the council, and the public.

B.5.2   SECTION 1504.2 CRITERIA FOR REFERRAL

Environmental referrals should be made to the council only after concerted, timely (as early as possible in the process), but unsuccessful attempts to resolve differences with the lead agency. In determining what environmental objections to the matter are appropriate to refer to the council, an agency should weigh potential adverse environmental impacts, considering:

a. Possible violation of national environmental standards or policies.
b. Severity.
c. Geographical scope.
d. Duration.
e. Importance as precedents.
f. Availability of environmentally preferable alternatives.

B.5.3   SECTION 1504.3 PROCEDURE FOR REFERRALS
AND RESPONSE

a. A federal agency making the referral to the council shall:
  1. Advise the lead agency at the earliest possible time that it intends to refer a matter to the council unless a satisfactory agreement is reached.
  2. Include such advice in the referring agency's comments on the draft environmental impact statement, except when the statement does not

contain adequate information to permit an assessment of the matter's environmental acceptability.

3. Identify any essential information that is lacking and request that it be made available at the earliest possible time.

4. Send copies of such advice to the council.

b. The referring agency shall deliver its referral to the council no later than 25 days after the final environmental impact statement has been made available to the Environmental Protection Agency, commenting agencies, and the public. Except when an extension of this period has been granted by the lead agency, the council will not accept a referral after that date.

c. The referral shall consist of:

1. A copy of the letter signed by the head of the referring agency and delivered to the lead agency informing the lead agency of the referral and the reasons for it, and requesting that no action be taken to implement the matter until the council acts upon the referral. The letter shall include a copy of the statement referred to in (c)(2) of this section.

2. A statement supported by factual evidence leading to the conclusion that the matter is unsatisfactory from the standpoint of public health or welfare or environmental quality. The statement shall:

    i. Identify any material facts in controversy and incorporate (by reference, if appropriate) the agreed upon facts.

    ii. Identify any existing environmental requirements or policies which would be violated by the matter.

    iii. Present the reasons why the referring agency believes the matter is environmentally unsatisfactory.

    iv. Contain a finding by the agency whether the issue raised is of national importance because of the threat to national environmental resources or policies or for some other reason.

    v. Review the steps taken by the referring agency to bring its concerns to the attention of the lead agency at the earliest possible time.

    vi. Give the referring agency's recommendations as to what mitigating alternative, further study, or other course of action (including abandonment of the matter) are necessary to remedy the situation.

d. No later than 25 days after referral to the council, the lead agency may deliver a response to the council and the referring agency. If the lead agency requests more time and gives assurance that the matter will not go forward in the interim, the council may grant an extension. The response shall:

1. Address fully the issues raised in the referral.

2. Be supported by the evidence.

3. Give the lead agency's response to the referring agency's recommendations.

e. Interested persons (including the applicant) may deliver their views in writing to the council. Views in support of the referral should be delivered no later than the referral. Views in support of the response shall be delivered no later than the response.

f. No later than 25 days after receipt of both the referral and any response or upon being informed that there will be no response (unless the lead agency agrees to a longer time), the council may take one or more of the following actions:

1. Conclude that the process of referral and response has successfully resolved the problem.
2. Initiate discussions with the agencies with the objective of mediation with the referring and lead agencies.
3. Hold public meetings or hearings to obtain additional views and information.
4. Determine that the issue is not one of national importance and request the referring and lead agencies to pursue their decision process.
5. Determine that the issue should be further negotiated by the referring and lead agencies and is not appropriate for council consideration until one or more heads of the agencies report to the council that the agencies' disagreements are irreconcilable.
6. Publish findings and recommendations (including where appropriate a finding that the submitted evidence does not support the position of an agency).
7. When appropriate, submit the referral and the response together with the council's recommendation to the president for action.

g. The council shall take no longer than 60 days to complete the actions specified in paragraph (f)(2), (3), or (5) of this section.

h. When the referral involves an action required by statute to be determined on the record after opportunity for agency hearing, the referral shall be conducted in a manner consistent with 5 U.S.C. 557(d) (Administrative Procedure Act).

[43 FR 55998, November 29, 1978; 44 FR 873, January 3, 1979].

B.6   PART 1505—NEPA AND AGENCY DECISION MAKING

Authority: NEPA, the Environmental Quality Improvement Act of 1970, as amended (42 U.S.C. 4371 *et seq.*), Section 309 of the Clean Air Act, as amended (42 U.S.C. 7609), and Executive Order 11514 (March 5, 1970, as amended by Executive Order 11991, May 24, 1977), from 43 FR 55999, November 29, 1978, unless otherwise noted.

B.6.1   SECTION 1505.1 AGENCY DECISION-MAKING PROCEDURES

Agencies shall adopt procedures (Section 1507.3) to ensure that decisions are made in accordance with the policies and purposes of the Act. Such procedures shall include but not be limited to:

a. Implementing procedures under Section 102(2) to achieve the requirements of Sections 101 and 102(1).

b. Designating the major decision points for the agency's principal programs likely to have a significant effect on the human environment and assuring that the NEPA process corresponds with them.

c. Requiring that relevant environmental documents, comments, and responses be part of the record in formal rule making or adjudicatory proceedings.

d. Requiring that relevant environmental documents, comments, and responses accompany the proposal through existing agency review processes so that agency officials use the statement in making decisions.

e. Requiring that the alternatives considered by the decision maker are encompassed by the range of alternatives discussed in the relevant environmental documents and that the decision maker consider the alternatives described in the environmental impact statement. If another decision document accompanies the relevant environmental documents to the decision maker, agencies are encouraged to make available to the public any part of that document that relates to the comparison of alternatives before the decision is made.

B.6.2   SECTION 1505.2 RECORD OF DECISION IN CASES REQUIRING ENVIRONMENTAL IMPACT STATEMENTS

At the time of its decision (Section 1506.10) or, if appropriate, its recommendation to congress, each agency shall prepare a concise public record of decision. The record, which may be integrated into any other record prepared by the agency, including that required by OMB Circular A-95 (revised), Part I, Sections 6(c) and (d), and Part II, Section 5(b)(4), shall:

a. State what the decision was.

b. Identify all alternatives considered by the agency in reaching its decision, specifying the alternative or alternatives that were considered to be environmentally preferable. An agency may discuss preferences among alternatives based on relevant factors including economic and technical considerations and agency statutory missions. An agency shall identify and discuss all such factors including any essential considerations of national policy which were balanced by the agency in making its decision and state how those considerations entered into its decision.

c. State whether all practicable means to avoid or minimize environmental harm from the alternatives selected have been adopted, and if not, why they were not. A monitoring and enforcement program shall be adopted and summarized where applicable for any mitigation.

B.6.3   SECTION 1505.3 IMPLEMENTING THE DECISION

Agencies may provide for monitoring to assure that their decisions are carried out and should do so in important cases. Mitigation [Section 1505.2(c)] and other conditions established in the environmental impact statement or during its review and committed as part of the decision shall be implemented by the lead agency or other appropriate consenting agency. The lead agency shall:

a. Include appropriate conditions in grants, permits, or other approvals.
b. Condition funding of actions on mitigation.
c. Upon request, inform cooperating or commenting agencies on the progress in carrying out the mitigating measures that they have proposed and that were adopted by the agency making the decision.
d. Upon request, make available to the public the results of relevant monitoring.

B.7   PART 1506—OTHER REQUIREMENTS OF NEPA

Authority: NEPA, the Environmental Quality Improvement Act of 1970, as amended (42 U.S.C. 4371 *et seq.*), Section 309 of the Clean Air Act, as amended (42 U.S.C. 7609), and Executive Order 11514 (March 5, 1970, as amended by Executive Order 11991, May 24, 1977), from 43 FR 56000, November 29, 1978, unless otherwise noted.

B.7.1   SECTION 1506.1 LIMITATIONS ON ACTIONS DURING NEPA PROCESS

a. Until an agency issues a record of decision as provided in Section 1505.2 [except as provided in paragraph (c) of this section], no action concerning the proposal shall be taken which would:
   1. Have an adverse environmental impact.
   2. Limit the choice of reasonable alternatives.
b. If any agency is considering an application from a nonfederal entity, and is aware that the applicant is about to take an action within the agency's jurisdiction that would meet either of the criteria in paragraph (a) of this section, then the agency shall promptly notify the applicant that the agency will take appropriate action to ensure that the objectives and procedures of NEPA are achieved.
c. While work on a required program environmental impact statement is in progress and the action is not covered by an existing program statement, agencies shall not undertake in the interim any major federal action covered by the program that may significantly affect the quality of the human environment unless such action:
   1. Is justified independently of the program.
   2. Is itself accompanied by an adequate environmental impact statement.
   3. Will not prejudice the ultimate decision on the program. Interim action prejudices the ultimate decision on the program when it tends to determine subsequent development or limit alternatives.
d. This section does not preclude development by applicants of plans or designs or performance of other work necessary to support an application

for federal, state, or local permits or assistance. Nothing in this section shall preclude Rural Electrification Administration approval of minimal expenditures not affecting the environment (e.g., long lead-time equipment and purchase options) made by non-governmental entities seeking loan guarantees from the administration.

B.7.2   SECTION 1506.2 ELIMINATION OF DUPLICATION WITH STATE AND LOCAL PROCEDURES

a. Agencies authorized by law to cooperate with state agencies of statewide jurisdiction pursuant to Section 102(2)(D) of the Act may do so.

b. Agencies shall cooperate with state and local agencies to the fullest extent possible to reduce duplication between NEPA and state and local requirements, unless the agencies are specifically barred from doing so by some other law. Except for cases covered by paragraph (a) of this section, such cooperation shall to the fullest extent possible include:
   1. Joint planning processes.
   2. Joint environmental research and studies.
   3. Joint public hearings (except where otherwise provided by statute).
   4. Joint environmental assessments.

c. Agencies shall cooperate with state and local agencies to the fullest extent possible to reduce duplication between NEPA and comparable state and local requirements, unless the agencies are specifically barred from doing so by some other law. Except for cases covered by paragraph (a) of this section, such cooperation shall to the fullest extent possible include joint environmental impact statements. In such cases, one or more federal agencies and one or more state or local agencies shall be joint lead agencies. Where state laws or local ordinances have environmental impact statement requirements in addition to but not in conflict with those in NEPA, federal agencies shall cooperate in fulfilling these requirements as well as those of federal laws so that one document will comply with all applicable laws.

d. To better integrate environmental impact statements into state or local planning processes, statements shall discuss any inconsistency of a proposed action with any approved state or local plan and laws (whether or not federally sanctioned). Where an inconsistency exists, the statement should describe the extent to which the agency would reconcile its proposed action with the plan or law.

B.7.3   SECTION 1506.3 ADOPTION

a. An agency may adopt a federal draft or final environment al impact statement or portion thereof provided that the statement or portion thereof meets the standards for an adequate statement under these regulations.

b. If the actions covered by the original environmental impact statement and the proposed action are substantially the same, the agency adopting another agency's statement is not required to recirculate it except as a final statement. Otherwise the adopting agency shall treat the statement as a draft and recirculate it [except as provided in paragraph (c) of this section].

c. A cooperating agency may adopt without recirculating the environmental impact statement of a lead agency when, after an independent review of the statement, the cooperating agency concludes that its comments and suggestions have been satisfied.

d. When an agency adopts a statement that is not final within the agency that prepared it, or when the action it assesses is the subject of a referral under Part 1504, or when the statement's adequacy is the subject of a judicial action that is not final, the agency shall so specify.

B.7.4   SECTION 1506.4 COMBINING DOCUMENTS

Any environmental document in compliance with NEPA may be combined with any other agency document to reduce duplication and paperwork.

B.7.5   SECTION 1506.5 AGENCY RESPONSIBILITY

a. Information—If an agency requires an applicant to submit environmetal information for possible use by the agency in preparing an environmental impact statement, then the agency should assist the applicant by outlining the types of information required. The agency shall independently evaluate the information submitted and shall be responsible for its accuracy. If the agency chooses to use the information submitted by the applicant in the environmental impact statement, either directly or by reference, then the names of the persons responsible for the independent evaluation shall be included in the list of preparers (Section 1502.17). It is the intent of this paragraph that acceptable work not be redone but that it be verified by the agency.

b. Environmental assessments—If an agency permits an applicant to prepare an environmental assessment, the agency, besides fulfilling the requirements of paragraph (a) of this section, shall make its own evaluation of the environmental issues and take responsibility for the scope and content of the environmental assessment.

c. Environmental impact statements—Except as provided in Sections 1506.2 and 1506.3, any environmental impact statement prepared pursuant to the requirements of NEPA shall be prepared directly by or by a contractor selected by the lead agency or where appropriate under Section 1501.6(b), a cooperating agency. It is the intent of these regulations that the contractor be chosen solely by the lead agency, or by the lead agency in cooperation with cooperating agencies, or where appropriate by a cooperating agency to avoid any conflict of interest. Contractors shall execute a disclosure statement prepared by the lead agency, or where appropriate the cooperating agency, specifying that they have no financial or other interest in the outcome of the project. If the document is prepared by contract, the responsible federal official shall furnish guidance and participate in the preparation and shall independently evaluate the statement prior to its approval and take responsibility for its scope and contents. Nothing in this section is intended to prohibit any agency from requesting any person to submit information to it or to prohibit any person from submitting information to any agency.

B.7.6 SECTION 1506.6 PUBLIC INVOLVEMENT

Agencies shall:

a. Make diligent efforts to involve the public in preparing and implementing their NEPA procedures.
b. Provide public notice of NEPA-related hearings, public meetings, and the availability of environmental documents so as to inform those persons and agencies who may be interested or affected.
   1. In all cases, the agency shall mail notice to those who have requested it on an individual action.
   2. In the case of an action with effects of national concern, notice shall include publication in the *Federal Register* and notice by mail to national organizations reasonably expected to be interested in the matter and may include listing in the 102 Monitor. An agency engaged in rule making may provide notice by mail to national organizations who have requested that notice regularly be provided. Agencies shall maintain a list of such organizations.
   3. In the case of an action with effects primarily of local concern the notice may include:
      i. Notice to state and area-wide clearinghouses pursuant to OMB Circular A-95 (revised).
      ii. Notice to Indian tribes when effects may occur on reservations.
      iii. Following the affected state's public notice procedures for comparable actions.
      iv. Publication in local newspapers (in papers of general circulation rather than legal papers).
      v. Notice through other local media.
      vi. Notice to potentially interested community organizations including small business associations.
      vii. Publication in newsletters that may be expected to reach potentially interested persons.
      viii. Direct mailing to owners and occupants of nearby or affected property.

      ix. Posting of notice on and off site in the area where the action is to be located.

  c. Hold or sponsor public hearings or public meetings whenever appropriate or in accordance with statutory requirements applicable to the agency. Criteria shall include whether there is:

    1. Substantial environmental controversy concerning the proposed action or substantial interest in holding the hearing.

    2. A request for a hearing by another agency with jurisdiction over the action supported by reasons why a hearing will be helpful. If a draft environmental impact statement is to be considered at a public hearing, the agency should make the statement available to the public at least 15 days in advance (unless the purpose of the hearing is to provide information for the draft environmental impact statement).

  d. Solicit appropriate information from the public.

  e. Explain in its procedures where interested persons can get information or status reports on environmental impact statements and other elements of the NEPA process.

  f. Make environmental impact statements, the comments received, and any underlying documents available to the public pursuant to the provisions of the Freedom of Information Act (5 U.S.C. 552) without regard to the exclusion for interagency memoranda where such memoranda transmit comments of federal agencies on the environmental impact of the proposed action. Materials to be made available to the public shall be provided to the public without charge to the extent practicable, or at a fee that is not more than the actual costs of reproducing copies required to be sent to other federal agencies, including the council.

B.7.7   SECTION 1506.7 FURTHER GUIDANCE

The council may provide further guidance concerning NEPA and its procedures including:

  a. A handbook which the council may supplement from time to time, which shall in plain language provide guidance and instructions concerning the application of NEPA and these regulations.

  b. Publication of the council's Memoranda to Heads of Agencies.

  c. In conjunction with the Environmental Protection Agency and the publication of the 102 Monitor, notice of:

    1. Research activities.

    2. Meetings and conferences related to NEPA.

    3. Successful and innovative procedures used by agencies to implement NEPA.

B.7.8   SECTION 1506.8 PROPOSALS FOR LEGISLATION

  a. The NEPA process for proposals for legislation (Section 1508.17) significantly affecting the quality of the human environment shall be integrated with the legislative process of the congress. A legislative environmental impact statement is the detailed statement required by law to be included in a recommendation or report on a legislative proposal to congress. A legislative environmental impact statement shall be considered part of the formal transmittal of a legislative proposal to congress; however, it may be transmitted to congress up to 30 days later in order to allow time for completion of an accurate statement that can serve as the basis for public and congressional debate. The statement must be available in time for congressional hearings and deliberations.

b. Preparation of a legislative environmental impact statement shall conform to the requirements of these regulations except as follows:

1. There need not be a scoping process.

2. The legislative statement shall be prepared in the same manner as a draft statement, but shall be considered the "detailed statement" required by statute; provided that when any of the following conditions exist both the draft and final environmental impact statement on the legislative proposal shall be prepared and circulated as provided by Sections 1503.1 and 1506.10.

   i. A congressional committee with jurisdiction over the proposal has a rule requiring both draft and final environmental impact statements.

   ii. The proposal results from a study process required by statute (such as those required by the Wild and Scenic Rivers Act [16 U.S.C. 1271 et seq.) and the Wilderness Act (16 U.S.C. 1131 et seq.)].

   iii. Legislative approval is sought for federal or federally assisted construction or other projects which the agency recommends be located at specific geographic locations. For proposals requiring an environmental impact statement for the acquisition of space by the General Services Administration, a draft statement shall accompany the prospectus or the 11(b) Report of Building Project Surveys to the congress, and a final statement shall be completed before site acquisition.

   iv. The agency decides to prepare draft and final statements.

c. Comments on the legislative statement shall be given to the lead agency which shall forward them along with its own responses to the congressional committees with jurisdiction.

B.7.9   SECTION 1506.9 FILING REQUIREMENTS

Environmental impact statements together with comments and responses shall be filed with the Environmental Protection Agency, attention Office of Federal Activities (A-104), 401 M Street SW., Washington, D.C. 20460. Statements shall be filed with the EPA no earlier than they are also transmitted to commenting agencies and made available to the public. The EPA shall deliver one copy of each statement to the council, which shall satisfy the requirement of availability to the president. The EPA may issue guidelines to agencies to implement its responsibilities under this section and Section 1506.10.

B.7.10   SECTION 1506.10 TIMING OF AGENCY ACTION

a. The Environmental Protection Agency shall publish a notice in the *Federal Register* each week of the environmental impact statements filed during the preceding week. The minimum time periods set forth in this section shall be calculated from the date of publication of this notice.

b. No decision on the proposed action shall be made or recorded under Section 1505.2 by a federal agency until the later of the following dates:

   1. 90 days after publication of the notice described above in paragraph (a) of this section for a draft environmental impact statement.

   2. 30 days after publication of the notice described above in paragraph (a) of this section for a final environmental impact statement. An exception to the rules on timing may be made in the case of an agency decision that is subject to a formal internal appeal. Some agencies have a formally established appeal process which allows other agencies or the public to take appeals on a decision and to make their views known after the publication of the final environmental impact statement. In such cases, where a real opportunity exists to alter the decision, the decision may be made and recorded at the same time the environmental impact statement is published.

   This means that the period for appeal of the decision and the 30 day period prescribed in paragraph (b)(2) of this section may run concurrently. In such cases the environmental impact statement shall explain the timing and the public's right of appeal. An agency engaged in rule making under the Administrative Procedure Act or other statute for the purpose of protecting the public health or safety may waive the time period in paragraph (b)(2) of this section and publish a decision on the final rule simultaneously with publication of the notice of the availability of the final environmental impact statement as described in paragraph (a) of this section.

c. If the final environmental impact statement is filed within 90 days after a draft environmental impact statement is filed with the Environmental Protection Agency, the minimum 30 day period and the minimum 90 day period may run concurrently. However, subject to paragraph (d) of this section, agencies shall allow not less than 45 days for comments on draft statements.

d. The lead agency may extend prescribed periods. The Environmental Protection Agency may upon a showing by the lead agency of compelling reasons of national policy reduce the prescribed periods and may upon a showing by any other federal agency of compelling reasons of national policy also extend prescribed periods, but only after consultation with the lead agency. [Also see Section 1507.3(d).] Failure to file timely comments shall not be a sufficient reason for extending a period. If the lead agency does not concur with the extension of time, the EPA may not extend it for more than 30 days. When the Environmental Protection Agency reduces or extends any period of time, it shall notify the council.

[43 FR 56000, November 29, 1978; 44 FR 874, January 3, 1979].

B.7.11   SECTION 1506.11 EMERGENCIES

Where emergency circumstances make it necessary to take an action with significant environmental impact without observing the provisions of these regulations, the federal agency taking the action should consult with the council about alternative arrangements. Agencies and the council will limit such arrangements to actions necessary to control the immediate impacts of the emergency. Other actions remain subject to NEPA review.

B.7.12   SECTION 1506.12 EFFECTIVE DATE

The effective date of these regulations is July 30, 1979, except that for agencies that administer programs that qualify under Section 102(2)(D) of the Act or under Section 104(h) of the Housing and Community Development Act of 1974, an additional four months shall be allowed for the state or local agencies to adopt their implementing procedures.

    a. These regulations shall apply to the fullest extent practicable to ongoing activities and environmental documents begun before the effective date. These regulations do not apply to an environmental impact statement or supplement if the draft statement was filed before the effective date of these regulations. No completed environmental documents need to be redone by reason of these regulations. Until these regulations are applicable, the council's guidelines published in the *Federal Register* of August 1, 1973, shall continue to be applicable. In cases where these regulations are applicable, the guidelines are superseded. However, nothing shall prevent an agency from proceeding under these regulations at an earlier time.

    b. NEPA shall continue to be applicable to actions begun before January 1, 1970, to the fullest extent possible.

B.8   PART 1507—AGENCY COMPLIANCE

Authority: NEPA, the Environmental Quality Improvement Act of 1970, as amended (42 U.S.C. 4371 *et seq.*), Section 309 of the Clean Air Act, as amended (42 U.S.C. 7609), and Executive Order 11514 (March 5, 1970, as amended by Executive Order 11991, May 24, 1977), from 43 FR 56002, November 29, 1978, unless otherwise noted.

B.8.1   SECTION 1507.1 COMPLIANCE

All agencies of the federal government shall comply with these regulations. It is the intent of these regulations to allow each agency flexibility in adapting its implementing procedures authorized by Section 1507.3 to the requirements of other applicable laws.

B.8.2   SECTION 1507.2 AGENCY CAPABILITY TO COMPLY

Each agency shall be capable (in terms of personnel and other resources) of complying with the requirements enumerated below. Such compliance may include use of other's resources, but the using agency shall itself have sufficient capability to evaluate what others do for it. Agencies shall:

a. Fulfill the requirements of Section 102(2)(A) of the Act to utilize a systematic, interdisciplinary approach which will ensure the integrated use of the natural and social sciences and the environmental design arts in planning and in decision making that may have an impact on the human environment. Agencies shall designate a person to be responsible for overall review of agency NEPA compliance.
b. Identify methods and procedures required by Section 102(2)(B) to ensure that presently unquantified environmental amenities and values may be given appropriate consideration.
c. Prepare adequate environmental impact statements pursuant to Section 102(2)(C) and comment on statements in the areas where the agency has jurisdiction by law or special expertise or is authorized to develop and enforce environmental standards.
d. Study, develop, and describe alternatives to recommended courses of action in any proposal which involves unresolved conflicts concerning alternative uses of available resources. This requirement of Section 102(2)(E) extends to all such proposals, not just the more limited scope of Section 102(2)(C)(iii) where the discussion of alternatives is confined to impact statements.
e. Comply with the requirements of Section 102(2)(H) that the agency initiate and utilize ecological information in the planning and development of resource-oriented projects.
f. Fulfill the requirements of Sections 102(2)(F), 102(2)(G), and 102(2)(I), of the Act and of Executive Order 11514, Protection and Enhancement of Environmental Quality, Section 2.

B.8.3   SECTION 1507.3 AGENCY PROCEDURES

a. Not later than eight months after publication of these regulations as finally adopted in the *Federal Register,* or five months after the establishment of an agency, whichever shall come later, each agency shall as necessary adopt procedures to supplement these regulations. When the agency is a department, major subunits are encouraged (with the consent of the department) to adopt their own procedures. Such procedures shall not paraphrase these regulations. They shall confine themselves to implementing procedures. Each agency shall consult with the council while developing its procedures and before publishing them in the *Federal Register* for comment. Agencies with similar programs should consult with each other and the council to coordinate their procedures, especially for programs requesting similar information from applicants. The procedures shall be adopted only after an opportunity for public review and after review by the council for conformity with the Act and these regulations.

The council shall complete its review within 30 days. Once in effect, they shall be filed with the council and made readily available to the public. Agencies are encouraged to publish explanatory guidance for these regulations and their own procedures. Agencies shall continue to review their policies and procedures and, in consultation with the council, to revise them as necessary to ensure full compliance with the purposes and provisions of the Act.

b. Agency procedures shall comply with these regulations except where compliance would be inconsistent with statutory requirements and shall include:

1. Those procedures required by Sections 1501.2(d), 1502.9(c)(3), 1505.1, 1506.6(e), and 1508.4.

2. Specific criteria for and identification of those typical classes of action:
    i. Which normally do require environmental impact statements.
    ii. Which normally do not require either an environmental impact statement or an environmental assessment [categorical exclusions (Section 1508.4)].
    iii. Which normally require environmental assessments but not necessarily environmental impact statements.

c. Agency procedures may include specific criteria for providing limited exceptions to the provisions of these regulations for classified proposals. They are proposed actions that are specifically authorized under criteria established by an Executive Order or statute to be kept secret in the interest of national defense or foreign policy and are in fact properly classified pursuant to such Executive Order or statute. Environmental assessments and environmental impact statements which address classified proposals may be safeguarded and restricted from public dissemination in accordance with the agencies' own regulations applicable to classified information. These documents may be organized so that classified portions can be included as annexes so that the unclassified portions can be made available to the public.

d. Agency procedures may provide for periods of time other than those presented in Section 1506.10 when necessary to comply with other specific statutory requirements.

e. Agency procedures may provide that where there is a lengthy period between the agency's decision to prepare an environmental impact statement and the time of actual preparation, the notice of intent required by Section 1501.7 may be published at a reasonable time in advance of preparation of the draft statement.

B.9   PART 1508—TERMINOLOGY AND INDEX

Authority: NEPA, the Environmental Quality Improvement Act of 1970, as amended (42 U.S.C. 4371 *et seq.*), Section 309 of the Clean Air Act, as amended (42 U.S.C. 7609), and Executive Order 11514 (March 5, 1970, as amended by Executive Order 11991, May 24, 1977), from 43 FR 56003, November 29, 1978, unless otherwise noted.

B.9.1   SECTION 1508.1 TERMINOLOGY

The terminology of this part shall be uniform throughout the federal government.

B.9.2   SECTION 1508.2 ACT

Act means the National Environmental Policy Act, as amended (42 U.S.C. 4321 *et seq.*) which is also referred to as "NEPA."

B.9.3   SECTION 1508.3 AFFECTING

Affecting means will or may have an effect on.

B.9.4   SECTION 1508.4 CATEGORICAL EXCLUSION

Categorical exclusion means a category of actions that do not individually or cumulatively have a significant effect on the human environment and that have been found to have no such effect in procedures adopted by a federal agency in implementation of these regulations (Section 1507.3) and for which, therefore, neither an environmental assessment nor an environmental impact statement is required. An agency may decide in its procedures or otherwise, to prepare environmental assessments for the reasons stated in Section 1508.9 even though it is not required to do so. Any procedures under this section shall provide for extraordinary circumstances in which a normally excluded action may have a significant environmental effect.

B.9.5   SECTION 1508.5 COOPERATING AGENCY

Cooperating agency means any federal agency other than a lead agency that has jurisdiction by law or special expertise with respect to any environmental impact involved in a proposal (or a reasonable alternative) for legislation or other major federal action significantly affecting the quality of the human environment. The selection and responsibilities of a cooperating agency are described in Section 1501.6. A state or local agency of similar qualifications or, when the effects are on a reservation, an Indian tribe, may by agreement with the lead agency become a cooperating agency.

B.9.6   SECTION 1508.6 COUNCIL

Council means the Council on Environmental Quality established by Title II of the Act.

B.9.7   SECTION 1508.7 CUMULATIVE IMPACT

Cumulative impact is the impact on the environment that results from the incremental impact of the action when added to other past, present, and reasonably foreseeable future actions regardless of what agency (federal or nonfederal) or person undertakes such other actions. Cumulative impacts can result from individually minor but collectively significant actions taking place over a period of time.

B.9.8   SECTION 1508.8 EFFECTS

Effects include:

a. Direct effects that are caused by the action and occur at the same time and place.
b. Indirect effects that are caused by the action and are later in time or farther removed in distance, but are still reasonably foreseeable. Indirect effects may include growth inducing effects and other effects related to induced changes in the pattern of land use, population density, or growth rate, and related effects on air and water and other natural systems, including ecosystems.

Effects and impacts as used in these regulations are synonymous. Effects includes ecological (such as the effects on natural resources and on the components, structures, and functioning of affected ecosystems), aesthetic, historic, cultural, economic, social, or health, whether direct, indirect, or cumulative. Effects may also include those resulting from actions that may have both beneficial and detrimental effects, even if on balance the agency believes that the effect will be beneficial.

B.9.9   SECTION 1508.9 ENVIRONMENTAL ASSESSMENT

Environmental assessment:

a. Means a concise public document, for which a federal agency is responsible, that serves to:
   1. Briefly provide sufficient evidence and analysis for determining whether to prepare an environmental impact statement or a finding of no significant impact.
   2. Aid an agency's compliance with the Act when no environmental impact statement is necessary.
   3. Facilitate preparation of a statement when one is necessary.
b. Shall include brief discussions of the need for the proposal, of alternatives as required by Section 102(2)(E), of the environmental impacts of the proposed action and alternatives, and a listing of agencies and persons consulted.

B.9.10   SECTION 1508.10 ENVIRONMENTAL DOCUMENT

Environmental document includes the documents specified in Section 1508.9 (environmental assessment), Section 1508.11 (environmental impact statement), Section 1508.13 (finding of no significant impact), and Section 1508.22 (notice of intent).

B.9.11   SECTION 1508.11 ENVIRONMENTAL IMPACT STATEMENT

Environmental impact statement means a detailed written statement as required by Section 102(2)(C) of the Act.

B.9.12   SECTION 1508.12 FEDERAL AGENCY

Federal agency means all agencies of the federal government. It does not mean the congress, the judiciary, or the president, including the performance of staff functions for the president in his executive office. It also includes, for purposes of these regulations, states and units of general local government and Indian tribes assuming NEPA responsibilities under Section 104(h) of the Housing and Community Development Act of 1974.

B.9.13   SECTION 1508.13 FINDING OF
NO SIGNIFICANT IMPACT

Finding of no significant impact means a document by a federal agency briefly presenting the reasons why an action, not otherwise excluded (Section 1508.4), will not have a significant effect on the human environment and for which an environmental impact statement therefore will not be prepared. It shall include the environmental assessment or a summary of it and shall note any other environmental documents related to it [Section 1501.7(a)(5)]. If the assessment is included, the finding need not repeat any of the discussion in the assessment but may incorporate it by reference.

B.9.14   SECTION 1508.14 HUMAN ENVIRONMENT

Human environment shall be interpreted comprehensively to include the natural and physical environment and the relationship of people with that environment. [See the definition of effects (Section 1508.8).] This means that economic or social effects are not intended by themselves to require preparation of an environmental impact statement. When an environmental impact statement is prepared and economic or social and natural or physical environmental effects are interrelated, then the environmental impact statement will discuss all of these effects on the human environment.

B.9.15   SECTION 1508.15 JURISDICTION BY LAW

Jurisdiction by law means agency authority to approve, veto, or finance all or part of the proposal.

B.9.16   SECTION 1508.16 LEAD AGENCY

Lead agency means the agency or agencies preparing or having taken primary responsibility for preparing the environmental impact statement.

B.9.17   SECTION 1508.17 LEGISLATION

Legislation includes a bill or legislative proposal to congress developed by or with the significant cooperation and support of a federal agency, but does not include requests for appropriations. The test for significant cooperation is whether the proposal is in fact predominantly that of the agency rather than another source. Drafting does not by itself constitute significant cooperation. Proposals for legislation include requests for ratification of treaties. Only the agency which has primary responsibility for the subject matter involved will prepare a legislative environmental impact statement.

B.9.18 Section 1508.18 Major Federal Action

Major federal action includes actions with effects that may be major and that are potentially subject to federal control and responsibility, reinforces, but does not have a significant independent meaning (Section 1508.27). Actions include the circumstance where the responsible officials fail to act and that failure to act is reviewable by courts or administrative tribunals under the Administrative Procedure Act or other applicable law as agency action.

a. Actions include new and continuing activities, including projects and programs entirely or partly financed, assisted, conducted, regulated, or approved by federal agencies; new or revised agency rules, regulations, plans, policies, or procedures; and legislative proposals (Sections 1506.8 and 1508.17). Actions do not include funding assistance solely in the form of general revenue sharing funds, distributed under the State and Local Fiscal Assistance Act of 1972, 31 U.S.C. 1221 *et seq.*, with no federal agency control over the subsequent use of such funds. Actions do not include bringing judicial or administrative, civil, or criminal enforcement actions.

b. Federal actions tend to fail within one of the following categories:
   1. Adoption of official policy, such as rules, regulations, and interpretations adopted pursuant to the Administrative Procedure Act, 5 U.S.C. 551 *et seq.;* treaties and international conventions or agreements; formal documents establishing an agency's policies which will result in or substantially alter agency programs.
   2. Adoption of formal plans, such as official documents prepared or approved by federal agencies that guide or prescribe alternative uses of federal resources, upon which future agency actions will be based.
   3. Adoption of programs, such as a group of concerted actions to implement a specific policy or plan; systematic and connected agency decisions allocating agency resources to implement a specific statutory program or executive directive.
   4. Approval of specific projects, such as construction or management activities located in a defined geographic area. Projects include actions approved by permit or other regulatory decision as well as federal and federally assisted activities.

B.9.19 Section 1508.19 Matter

Matter includes for purposes of Part 1504:

a. With respect to the Environmental Protection Agency, any proposed legislation, project, action, or regulation as those terms are used in Section 309(a) of the Clean Air Act (42 U.S.C. 7609).

b. With respect to all other agencies, any proposed major federal action to which Section 102(2)(C) of NEPA applies.

B.9.20   SECTION 1508.20 MITIGATION

Mitigation includes:

a. Avoiding the impact altogether by not taking a certain action or parts of an action.
b. Minimizing impacts by limiting the degree or magnitude of the action and its implementation.
c. Rectifying the impact by repairing, rehabilitating, or restoring the affected environment.
d. Reducing or eliminating the impact over time by preservation and maintenance operations during the life of the action.
e. Compensating for the impact by replacing or providing substitute resources or environments.

B.9.21   SECTION 1508.21 NEPA PROCESS

NEPA process means all measures necessary for compliance with the requirements of Section 2 and Title I of NEPA.

B.9.22   SECTION 1508.22 NOTICE OF INTENT

Notice of intent means a notice that an environmental impact statement will be prepared and considered. The notice shall briefly:

a. Describe the proposed action and possible alternatives.
b. Describe the agency's proposed scoping process including whether, when, and where any scoping meeting will be held.
c. State the name and address of a person within the agency who can answer questions about the proposed action and the environmental impact statement.

B.9.23   SECTION 1508.23 PROPOSAL

Proposal exists at that stage in the development of an action when an agency subject to the Act has a goal and is actively preparing to make a decision on one or more alternative means of accomplishing that goal, and the effects can be meaningfully evaluated. Preparation of an environmental impact statement on a proposal should be timed (Section 1502.5) so that the final statement may be completed in time for the statement to be included in any recommendation or report on the proposal. A proposal may exist in fact as well as by agency declaration that one exists.

B.9.24   SECTION 1508.24 REFERRING AGENCY

Referring agency means the federal agency which has referred any matter to the council after a determination that the matter is unsatisfactory from the standpoint of public health or welfare or environmental quality.

B.9.25   SECTION 1508.25 SCOPE

Scope consists of the range of actions, alternatives, and impacts to be considered in an environmental impact statement. The scope of an individual statement may depend on its relationships to other statements (Sections 1502.20 and 1508.28). To determine the scope of environmental impact statements, agencies shall consider three types of actions, three types of alternatives, and three types of impacts. They include:

a. Actions (other than unconnected single actions) that may be:
   1. Connected actions, which means that they are closely related and therefore should be discussed in the same impact statement. Actions are connected if they:
      i. Automatically trigger other actions that may require environmental impact statements.
      ii. Cannot or will not proceed unless other actions are taken previously or simultaneously.
      iii. Are interdependent parts of a larger action and depend on the larger action for their justification.
   2. Cumulative actions that, when viewed with other proposed actions, have cumulatively significant impacts and should therefore be discussed in the same impact statement.
   3. Similar actions that, when viewed with other reasonably foreseeable or proposed agency actions, have similarities that provide a basis for evaluating their environmental consequences together, such as common timing or geography. An agency may wish to analyze these actions in the same impact statement. It should do so when the best way to assess adequately the combined impacts of similar actions or reasonable alternatives to such actions is to treat them in a single impact statement.
b. Alternatives that include:
   1. No action alternative.
   2. Other reasonable courses of actions.
   3. Mitigation measures (not in the proposed action).
c. Impacts that may be:
   1. Direct.
   2. Indirect.
   3. Cumulative.

B.9.26   SECTION 1508.26 SPECIAL EXPERTISE

Special expertise means statutory responsibility, agency mission, or related program experience.

B.9.27   SECTION 1508.27 SIGNIFICANTLY

Significantly as used in NEPA requires considerations of both context and intensity:

a. Context—this means that the significance of an action must be analyzed in several contexts such as society as a whole (human, national), the affected region, the affected interests, and the locality. Significance varies with the setting of the proposed action. For instance, in the case of a site-specific action, significance would usually depend upon the effects in the locale rather than in the world as a whole. Both short- and long-term effects are relevant.

b. Intensity—this refers to the severity of impact. Responsible officials must bear in mind that more than one agency may make decisions about partial aspects of a major action. The following should be considered in evaluating intensity:

   1. Impacts that may be both beneficial and adverse—a significant effect may exist even if the federal agency believes that on balance the effect will be beneficial.
   2. The degree to which the proposed action affects public health or safety.
   3. Unique characteristics of the geographic area such as proximity to historic or cultural resources, park lands, prime farmlands, wetlands, wild and scenic rivers, or ecologically critical areas.
   4. The degree to which the effects on the quality of the human environment are likely to be highly controversial.
   5. The degree to which the possible effects on the human environment are highly uncertain or involve unique or unknown risks.
   6. The degree to which the action may establish a precedent for future actions with significant effects or represents a decision in principle about a future consideration.
   7. Whether the action is related to other actions with individually insignificant but cumulatively significant impacts. Significance exists if it is reason to anticipate a cumulatively significant impact on the environment. Significance cannot be avoided by terming an action temporary or by breaking it down into small component parts.
   8. The degree to which the action may adversely affect districts, sites, highways, structures, or objects listed in or eligible for listing in the National Register of Historic Places or may cause loss or destruction of significant scientific, cultural, or historical resources.
   9. The degree to which the action may adversely affect an endangered or threatened species or its habitat that has been determined to be critical under the Endangered Species Act of 1973.
   10. Whether the action threatens a violation of federal, state, or local law or requirements imposed for the protection of the environment.

[43 FR 56003, November 29, 1978; 44 FR 874, January 3, 1979].

B.9.28   SECTION 1508.28 TIERING

Tiering refers to the coverage of general matters in broader environmental impact statements (such as national program or policy statements) with subsequent narrower statements or environmental analyses (such as regional or basin-wide program statements or ultimately site-specific statements) incorporating by reference the general discussions and concentrating solely on the issues specific to the statement subsequently prepared. Tiering is appropriate when the sequence of statements or analyses is:

a. From a program, plan, or policy environmental impact statement to a program, plan, or policy statement or analysis of lesser scope or to a site-specific statement or analysis.
b. From an environmental impact statement on a specific action at an early stage (such as need and site selection) to a supplement (that is preferred) or a subsequent statement or analysis at a later stage (such as environmental mitigation). Tiering in such cases is appropriate when it helps the lead agency to focus on the issues which are ripe for decision and exclude from consideration issues already decided or not yet ripe.

C Federal Agency Regulations Implementing NEPA

- Advisory Council on Historic Preservation 36 CFR 805 (1988).
- Agency for International Development 22 CFR 216 (1988).

C.1 DEPARTMENT OF AGRICULTURE

- Department of Agriculture 7 CFR 1b, 3100 (1988).
- Agricultural Stabilization and Conservation Service 7 CFR 799 (1988).
- Animal and Plant Health Inspection Service 44 FR 50381 (8/28/79).
- Forest Service, *Forest Service Manual,* Chapter 1950.
- 50 FR 26078 (6/24/85), as amended by:
  - 52 FR 30935 (8/18/87) 17 ELR 10407 10/87.
  - 54 FR 9073 (3/3/89) 19 ELR 10273 5/89.
- Soil Conservation Service 7 CFR 650 (1988).
- Rural Electrification Administration 7 CFR 1794 (1988).
- Arms Control Disarmament Agency 45 FR 69510 (10/21/80).
- Central Intelligence Agency 44 FR 45431 (8/2/79).

C.2 DEPARTMENT OF COMMERCE

- Economic Development Administration 45 FR 63310 (9/24/80), as amended by:
  - 45 FR 74902 (11/13/80) 10 ELR 10204 11/80.
- National Oceanic and Atmospheric Administration 48 FR 14734 (4/5/83).
- Consumer Product Safety Commission 16 CFR 1021 (1988).

C.3 DEPARTMENT OF DEFENSE

- Department of Defense 32 CFR 214 (1988).
- Department of the Air Force 32 CFR 214 (1988).
- Department of the Army 32 CFR 650 and 651 (1988), as amended by:
  - 53 FR 46322 (11/16/88) 19 ELR 10037 1/88.
- Army Corps of Engineers 33 CFR 230 (1988).
- Department of the Navy 32 CFR 775 (1988).
- Delaware River Basin Commission 18 CFR 401, Subpart D (1988).

C.4   DEPARTMENT OF ENERGY

- Department of Energy 45 FR 20694 (3/28/80), as amended by:
  - 47 FR 7976 (2/23/82) and 48 FR 685 (1/6/83).
  - 52 FR 659 (1/7/87) and 52 FR 47662 (12/15/87).
  - Reprinted in entirety
- Federal Energy Regulatory Commission 18 CFR 2.80 and 380 (1988).
- Environmental Protection Agency 40 CFR 6 (1988).
- Export–Import Bank 12 CFR 408 (1988).
- Federal Communications Commission 47 CFR 1, Subpart I (1988).
- Federal Emergency Management Agency 44 CFR 10 (1987).
- Federal Maritime Commission 46 CFR 504 (1988).
- Federal Trade Commission 16 CFR 1, Subpart I (1988).
- General Services Administration 50 FR 7648 (2/25/85).

C.5   DEPARTMENT OF HEALTH
AND HUMAN SERVICES

- Departmental 45 FR 76519 (11/19/80) as amended by:
  - 47 FR 2414 (1/15/82) 12 ELR 05021 (2/1982)
- Food and Drug Administration 21 CFR 25 (1988)

C.6   DEPARTMENT OF HOUSING
AND URBAN DEVELOPMENT

- Departmental 24 CFR 50 and 51 (1988) as amended by:
  - 53 FR 11224 (4/5/88).
- Community Development Block Grant Program 24 CFR 58 (1988). 53 FR 30186 (8/10/88).

C.7   DEPARTMENT OF THE INTERIOR

- Departmental Departmental Manual; Part 516
- 45 FR 27541 (4/23/80) as amended by:
  - 49 FR 21437 (5/21/84) 14 ELR 10286 7/84.
- Bureau of Indian Affairs 46 FR 7490 (1/23/81).
- Bureau of Land Management 46 FR 7492 (1/23/81) as amended by:
  - 48 FR 43731 (9/26/83) 13 ELR 10385 11/83.
- Bureau of Mines 45 FR 85528 (12/29/80).
- Bureau of Reclamation 45 FR 47944 (7/17/80) as amended by:
  - 48 FR 17151 (4/21/83).
- Fish and Wildlife Service 45 FR 47941 (7/17/80) as amended by:
  - 47 FR 28841 (7/1/82) 12 ELR 05095 8/82 and 49 FR 7881 (3/2/84) 14 ELR 10182 4/84.
- Geological Survey 46 FR 7485 (1/23/81).
- Minerals Management Service 51 FR 1855 (1/15/86).

- National Park Service 46 FR 1042 (1/5/81).
- International Boundary and Water Commission 46 FR 44083 (9/2/81).
- International Communication Agency 44 FR 45489 (8/2/79).
- Interstate Commerce Commission 49 CFR 1105 (1987), as amended by:
  - 54 FR 9822 (3/8/89) 19 ELR 10241 5/89.

C.8   DEPARTMENT OF JUSTICE

- Departmental 28 CFR 61 (1988).
- Bureau of Prisons 28 CFR 61, Appendix A (1988).
- Drug Enforcement Administration 28 CFR 61, Appendix B (1988).
- Immigration and Naturalization Service 28 CFR 61, Appendix C (1988).
- Office of Justice Assistance, Research, and Statistics 28 CFR 61, Appendix D (1988).
- Department of Labor 29 CFR 11 (1988).
- Marine Mammal Commission 50 CFR 530 (1988).
- National Aeronautics and Space Administration 14 CFR 1216 (1988) as amended by:
  - 53 FR 9759 (3/25/88).
- National Capital Planning Commission 44 FR 64923 (11/8/79), as amended by:
  - 46 FR 51327 (10/19/81) 11 ELR 10231 12/81.
  - 47 FR 51481 (11/15/82) 13 ELR 10028 1/83.
- National Science Foundation 45 CFR 640 (1987).
- Nuclear Regulatory Commission 10 CFR 51 (1988), as amended by:
  - 53 FR 13399 (4/25/88), 53 FR 24018 (6/27/88) 18 ELR 10321 8/88, and 53 FR 31651 (8/19/88) 18 ELR 10454 10/88.
- 54 FR 15372 (4/18/89) 19 ELR 10276 6/89, Overseas Private Investment Corporation 44 FR 51385 (8/31/79).
- Pennsylvania Avenue Development Corporation 34 CFR 907 (1988).
- Postal Service 39 CFR 775 (1988).
- Saint Lawrence Seaway Development Corporation 46 FR 28795 (5/28/81).
- Securities and Exchange Commission 17 CFR 200, Subpart K (1988).
- Small Business Administration 45 FR 7358 (2/1/80).
- Department of State 22 CFR 161 (1988).
- Tennessee Valley Authority 45 FR 54511 (8/15/80), as amended by:
  - 47 FR 54586 (12/2/82) 13 ELR 10029 1/83.
  - 48 FR 19264 (4/28/83) 13 ELR 10193 (6/83).

C.9   DEPARTMENT OF TRANSPORTATION

- Departmental 44 FR 56420 (10/1/79).
- Coast Guard 45 FR 32816 (5/19/80), as amended by:
  - 50 FR 32944 (8/15/85) 15 ELR 10307 (9/85).
- Federal Aviation Administration 45 FR 2244 (1/10/80), as amended by:
  - 49 FR 28501 (7/12/84) 14 ELR 10325 (8/84).

- Federal Highway Administration 23 CFR 771 (1988).
- Federal Railroad Administration 45 FR 40854 (6/16/80).
- National Highway Traffic Safety Administration 49 CFR 520 (1987).
- Urban Mass Transportation Administration 23 CFR 771 (1988).
- Department of the Treasury 45 FR 1828 (1/8/80).
- Veterans Administration 38 CFR 26 (1988).
- Water Resources Council 18 CFR 707 (1988).

Index